Electrical Installation Technology 1: Theory and Regulations

Maurice L. Lewis

B Ed (Hons), FIEIE

STANLEY THORNES (PUBLISHERS) LTD

Originally published in 1980 by Hutchinson Education
Second edition 1984
Third edition 1988

Reprinted 1989 by Stanley Thornes (Publishers) Ltd
Ellenborough House
Wellington Street
CHELTENHAM GL50 1YW
United Kingdom

Third Edition (revised) 1992

98 99 00 / 10 9 8 7 6 5

British Library Cataloguing in Publication Data
Lewis, M. L.
Electrical installation technology: 1. Theory and regulations. – 3rd revised ed.
I. Title
621.319

ISBN 0–7487–1542–8

Photoset in Linotron Times by
Northern Phototypesetting Co. Ltd, Bolton.
Printed and bound in Great Britain
by Athenæum Press Ltd, Gateshead, Tyne & Wear

Contents

Preface

This is a revised textbook for electrical installation work students pursuing the City and Guilds Course 236, Parts I and II Certificates and it will prove to be a good reference book for students studying other electrical courses, particularly City and Guilds Courses 185, 201 and 232. First year B/TEC National Certificate students studying building services will also find areas of the book extremely useful since electricity in buildings today is given high priority in terms of personal comfort and building management systems.

As a revised book, Chapter 2 on *health and safety* now contains the latest changes in statutory and non-statutory regulations, e.g. the Electricity at Work Regulations 1989 and the 16th Edition of the IEE Wiring Regulations 1991. These latter regulations now include a new numbering system which is used throughout other chapters in the book. The chapters on *installations* have both been completely revised and modifications have been made to the last chapter on *inspection and testing*.

Answers to exercise questions and references are given at the end of the book.

M. L. L.
1992

Acknowledgements

The author wishes to thank several manufacturers and organisations who have contributed information to make this book possible, particularly:

AEI Cables Ltd
Barton Conduits Ltd
Brooks Motors Ltd
Crompton Parkinson Ltd
Electrical Review
Evershed & Vignoles Ltd
J. A. Crabtree Ltd
Midland Electric Manufacturing Co. Ltd
MK Electric Ltd
Reyrolle Ltd
The British Standards Institution
The City and Guilds of London Institute
The Health and Safety Executive
The Institution of Electrical Engineers
The Joint Industry Board
Thorn Lighting Ltd

The extracts from HSE Guidance Notes (F2508) which appear on pages 28–34 are reproduced with permission from the Controller of Her Majesty's Stationery Office.

chapter one

Terminology

After reading this chapter you should be able to:

1 state a number of useful definitions associated with electrical terminology,

2 state the meaning of a number of common electrical abbreviations, as well as names of important bodies related to the electrical industry,

3 recognise and distinguish between various BS 3939 graphical location symbols and circuit diagram symbols,

4 draw BS 3939 symbols in different types of circuit.

It is most important for electrical apprentices to quickly acquire a firm understanding of the basic terminology so widely used in their industry. One of the main considerations in the review of the City and Guilds electrical installation scheme was that students should develop the ability to communicate their jobs to others; this can be most satisfying providing that the terminology used is correct as well as fully understood.

Examining bodies continue to report to colleges about students' poor exam performance and one often comes across a general comment which reads: 'The standard of this paper was below what should be expected from trade electricians, particularly where testing is concerned.' Other comments are: 'Many used BS location symbols instead of circuit symbols.' and 'A popular question but not well answered, due mainly to the students' limited powers of description.'

Students ought really to break away from the habit of using personal pronouns like 'I' and 'me', particularly when they are asked to provide a description. It is the dreadful misuse of terminology that really must be put right. All too frequently the wrong term is used; for example, 'The house was wired using two ring mains.' The student should be aware of the fact that he is describing two ring final circuits. Ring mains are primary distribution circuits used for supplies into premises. Another source of confusion is the term 'bulb', which is part of a lamp that itself is part of a luminaire, previously called a lighting fitting. Pitfalls often occur between terms, such as joint box and junction box and 'Megger' and mega. It is still surprising how many students cannot explain the difference between conductor resistance and insulation resistance.

The terminology used in this chapter is directed at installation theory and practice; it is by no means a complete coverage of all that has to be learnt in the electrical contracting industry. There are many trade catalogues and British Standards Institution leaflets available in most college libraries. As an introduction, students ought to look at Appendix 1 of the IEE Wiring Regulations, where a list of important documents can be found.

Useful definitions

Accessory: a device, other than current-using equipment, associated with such equipment or with the wiring of an installation. For example, a lampholder, plug, switch, etc. but not a motor or luminaire.

Adaptor: an accessory for insertion into a socket outlet.

Ambient temperature: the air temperature or other medium in which the equipment is to be used.

Apparatus: the machines, equipment and fittings in which conductors are used but not the conductors themselves.

Appliance: any device that is designed to use electricity for a particular purpose, excluding a luminaire or independent motor.

Arm's reach: a zone of accessibility to touch, extending from any point on a surface where persons usually stand or move about, to the limits of reach with the hands in any direction without assistance.

Barrier: a part intended to prevent contact with live parts.

Bonded: the connection of metal items to ensure that a common potential exists.

Buzzer: a device to provide audible warning.

Cable: one or more conductors provided with insulation.

Circuit: the path taken by current to supply electrical equipment or to allow leakage current to return.

Circuit breaker: a mechanical device designed to open or close a circuit under normal or abnormal conditions.

Conductor: the conducting part of a cable, busbar or functioning part of metalwork that carries current.

Connector: a device used for connecting together flexible cords and cables.

Consumer unit: a combined fuseboard and main switch controlling and protecting a consumer's final circuits. Such units may contain residual current devices for protection against electric shock.

Consumer terminal: the point at the origin of a consumer's installation where the incoming supply of energy is delivered.

Contactor: a power-controlling device having an operating coil and contacts used for opening and closing circuits.

Dead: at or about earth potential and disconnected from any live system.

Design current: the current intended to be carried by a circuit in normal service.

Direct contact: contact of persons or livestock with live parts, which may result in an electric shock. *Indirect contact* is the possibility of obtaining an electric shock from exposed conductive parts made live by a fault.

Distribution board: an assemblage of excess current protective devices in an enclosure, with the purpose of protecting final circuits.

Double-insulation: an insulating medium comprising both functional insulation and protective insulation. Class II equipment uses such insulation.

Duct: A closed passageway formed to incorporate cables, often as an underground wiring system.

Earth: the conductive mass of earth whose electric potential at any point is taken as zero.

Earth electrode: a conductor or group of conductors in intimate contact with and providing an electrical contact with the general mass of earth.

Earth fault loop impedance: the total opposition to current flow, starting and ending at the point of fault.

Earth leakage current: a current that flows to earth.

Earthed concentric wiring: a wiring system in which one or more insulated conductors are completely surrounded throughout their length by a metal sheath that acts as a PEN conductor.

Earthing conductor: a protective conductor that connects a main earthing terminal to an earth electrode or to other means of earthing.

Earthing terminal: the main earth connection point at the intake position.

Electrical installation: an assemblage of electrical apparatus and equipment to fulfil a specific purpose within a consumer's premises.

Enclosure: a part providing a degree of protection against contact with live parts.

Exposed conductive part: a part of equipment that can be touched and is likely to become live under fault conditions. An *extraneous conductive part* does not form part of the electrical installation but since it may be conductive it is liable to introduce a potential, i.e. earth potential.

Final circuit: a circuit connected directly to current-using equipment through outlet points.

Flexible cord: a flexible cable in which the cross-sectional area of each conductor does not exceed 4 mm^2.

Fuse: A device for opening a circuit by means of a fuse-element.

Heater: an electrical appliance such as a radiant heater, convector heater or even an immersion heater.

Instructed person: a person advised or supervised by skilled persons to enable him to avoid the dangers that electricity may create.

Insulation: suitable non-conducting material enclosing, surrounding and supporting a conductor.

Integrated meter: an instrument such as a kilowatt-hour meter used for recording the amount of energy used by an electrical installation.

Isolation: the cutting off of a circuit or circuits from the source of electrical energy. An *isolator* is a devise used for this purpose.

Joint box: a box forming part of a wiring installation,

provided to contain joints in the conductors of cables.
Junction box: a box used for connecting two or more lengths of conduit or trunking.
Live: a conductor or object is said to be live when a potential difference exists between it and earth.
Luminaire: a lighting fitting that includes all the necessary parts for supporting, fixing and protecting lamps, together with any necessary controlgear.
Motor: a machine that converts electrical power into mechanical power, acting as a drive.
Multimeter: a universal indicating instrument capable of measuring current, voltage and resistance, as well as other quantities.
Obstacle: a part preventing unintentional contact with live parts but not preventing deliberate contact.
Ohmmeter: an instrument for measuring resistance, such as conductor resistance and insulation resistance.
Overcurrent: a current exceeding the rated value. It may be an overload or a short circuit; the former occurs in a circuit that is electrically sound and the latter occurs in a circuit resulting from a fault between live conductors.
PEN conductor: a conductor combining the function of both protective conductor and neutral conductor.
Plug: a device intended for connection to a flexible cord or flexible cable.
Point (in wiring): any termination of the fixed wiring intended for the attachment of electrical equipment, e.g. a socket point or a lighting point.
Power factor meter: an indicating instrument for measuring the unity power factor, lagging power or leading power factor conditions of a circuit.
Protective conductor: a conductor used for some measure of protection against electric shock. The term encompasses circuit protective conductors, main equipotential bonding conductors, earthing conductors and supplementary equipotential bonding conductors.
Rectifier: a device for converting a.c. into d.c. by allowing the passage of current to flow in one direction.
Residual current device: a device intended to cause the opening of contacts when its trip mechanism attains a given value.
Resistance area (of an earth electrode): the area of ground within which a voltage gradient exists when the electrode is under test.
Ring final circuit: a final circuit arranged in the form of a ring and connected to a single point of the supply.
Skilled person: a person with technical knowledge or sufficient experience to enable him to avoid the dangers that electricity may create.
Space factor: the ratio (expressed as a percentage) of the sum of the overall cross-sectional areas of cables forming a bunch to the internal cross-sectional area of conduit or trunking.
Spur: a branch cable connected to a ring final circuit.
Switch: a mechanical device for making and breaking a circuit. A *linked switch* is one that is designed to break all poles of the supply simultaneously, and a *time switch* is one that embodies a clock to operate electrical contacts. There are many types of functional switch: the *dimmer switch* for adjusting levels of illuminance, the *single-pole, two-way* and *intermediate switch* for general use, as well as the *fireman's switch* used to control high-voltage signs.
Switchgear: apparatus for controlling the distribution of electrical energy.
System: an electrical system comprising a single source of electrical energy and an installation. Such a system has a direct relationship with the type of earthing arrangement provided for it. The common ones are known as: (i) *TN–S system*, (ii) *TN–C–S system* and (iii) *TT system*. The first utilises a separate neutral and separate earth return, such as the supply cable sheath/armouring. The second utilises a PEN conductor throughout the system and used to be known as protective multiple earthing. The third system is used when the consumer has to provide his installation with its own earth electrode, as in some cases when the supply is from overhead cables.
Thermostat: a device that provides automatic control of temperature.
Transformer: a static device in which electrical power is transferred from one winding (or windings) by means of electromagnetic induction, usually to achieve a change of voltage.
Trunking: a fabricated casing for cables. One of the common wiring systems used today.
Voltage: the source of supply pressure that causes current to flow. *Nominal voltage* is the voltage

designated to an electrical installation, e.g. 240 V or 415 V. There are two nominal voltage ranges, namely:
(i) *Extra-low voltage*, which normally does not exceed 50 V a.c. or 120 V d.c., whether between conductors or to earth.
(ii) *Low voltage*, which normally exceeds extra-low voltage but does not exceed 1000 V a.c. or 1500 V d.c. between conductors, or 600 V a.c. or 900 V d.c. between conductors and earth.

Wattmeter: an indicating instrument for measuring electrical power.

Common abbreviations

Term	*Meaning*
a.c.	alternating current
c.p.c.	circuit protective conductor
c.t.	current transformer
d.c.	direct current
h.b.c.	high breaking capacity
h.o.f.r.	heat/oil resisting and flame retardant
m.i.m.s.	mineral-insulated metal sheath
p.c.p.	polychloroprene
p.i.l.c.s.a.	paper-insulated lead-covered steel armouring
p.m.e.	protective multiple earthing
r.c.d.	residual current device
x.l.p.e.	cross-linked polyethylene
PE	protective conductor
PVC	polyvinyl chloride
SP	single-pole
SPN	single-pole and neutral
TP	triple-pole
TPN	triple-pole and neutral

Names of important bodies

ASEE	Association of Supervisory and Executive Engineers
BSI	British Standards Institution
CENELEC	European Committee for Electrotechnical Standardisation
CGLI	City and Guilds of London Institute
ECA	Electrical Contractors' Association
EETPU	Electrical, Electronic, Telecommunication and Plumbing Trades Union
HSE	Health and Safety Executive
IEC	International Electrotechnical Commission
IEE	Institution of Electrical Engineers
IEEIE	Institution of Electrical and Electronics Incorporated Engineers
JIB	Joint Industry Board
JTL	JT Ltd.
NICEIC	National Inspection Council for Electrical Installation Contracting
REC	Regional Electricity Company
TEC	Technician Education Council

Conductor polarities

Conductor name	*Colour identification*
Non-flexible cables or bare conductors for fixed wiring	
3-phase, 4-wire a.c. circuits:	
R-phase (L1)	red
Y-phase (L2)	yellow
B-phase (L3)	blue
Neutral	black
Protective	green/yellow
1-phase, 2-wire a.c. circuits:	
Phase	red
Neutral	black
Protective	green/yellow
3-wire d.c. circuits:	
Positive (+)	red
Negative (−)	blue
Middle	black
Protective	green/yellow
2-wire d.c. circuits:	
Positive (+)	red
Negative (−)	black
Protective	green/yellow
Flexible cables and cords	
3-phase, 4-wire a.c. circuits:	
R/Y/B-phases	brown
Neutral	blue
Protective	green/yellow
1-phase, 2-wire a.c. circuits:	
Phase	brown
Neutral	blue
Protective	green/yellow

Drawing symbols

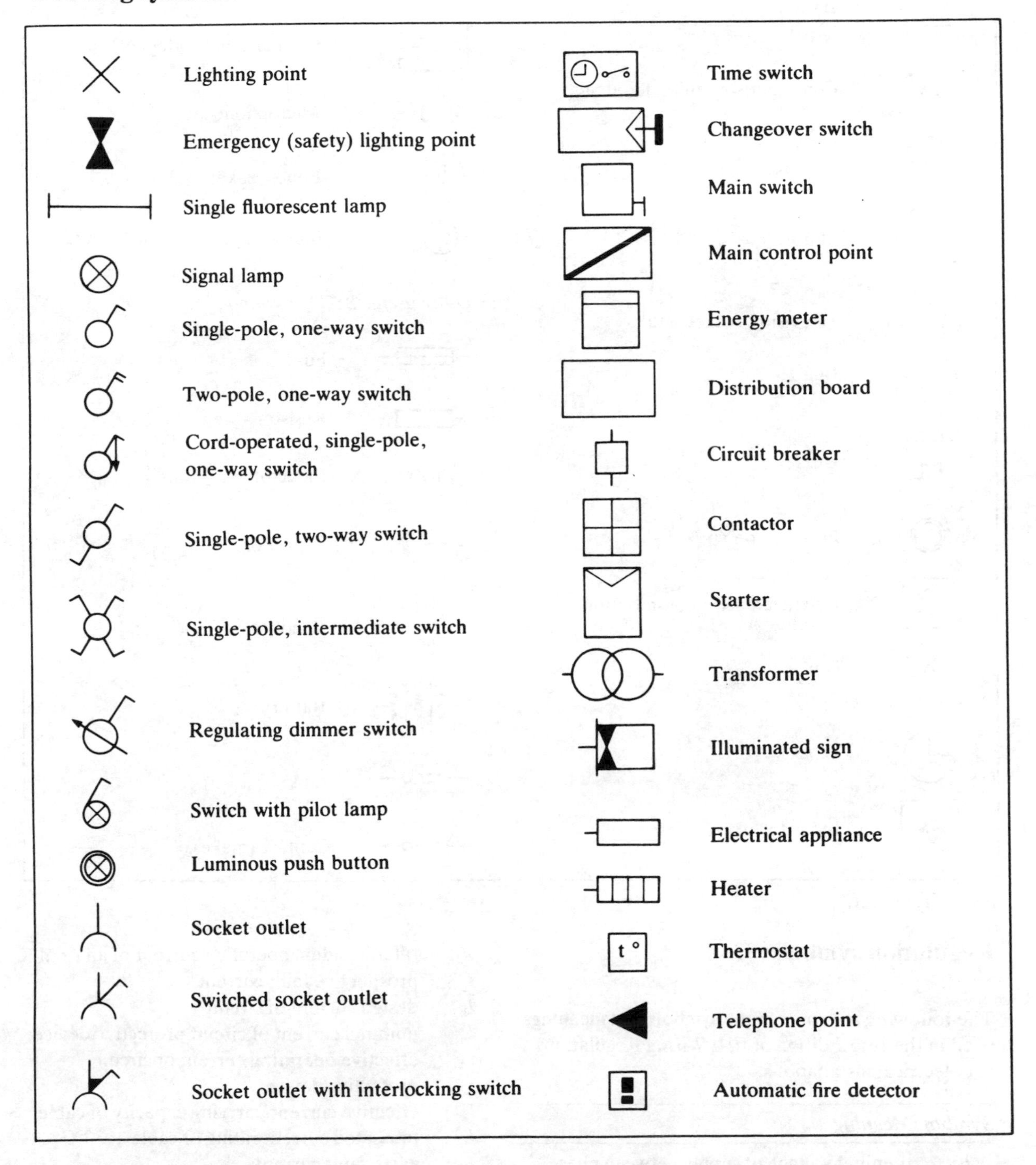

Figure 1.1 *BS 3939 Graphical symbols for architectural and installation diagrams*

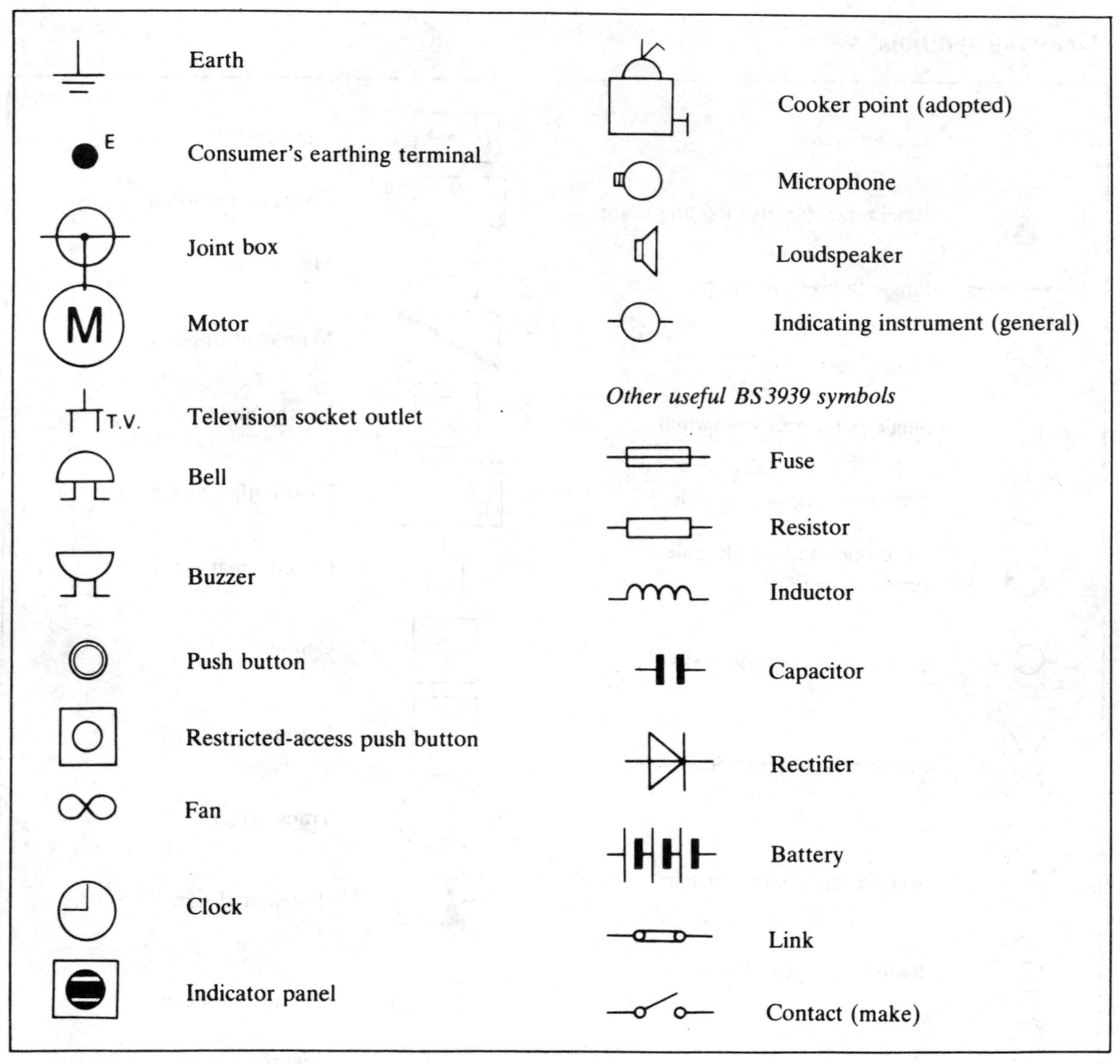

Figure 1.1 (cont.)

Regulation symbols

The following are some of the symbols and meanings used in the 16th Edition of IEE Wiring Regulations for electrical installations.

Symbol	*Meaning*
U_0	nominal voltage of supply between phase and neutral
I_a	operating current of a protective device
$I_{\Delta a}$	rated residual operating current of an r.c.d.
I_p	prospective fault current
I_b	design current of circuit
I_n	nominal current of circuit protective device
I_2	effective operating current of circuit protective device
I_z	effective current-carrying capacity of cable
I_t	tabulated current rating of cable at 30°C
I_F	earth fault current
C	correction factor
C_a	correction factor for ambient temperature

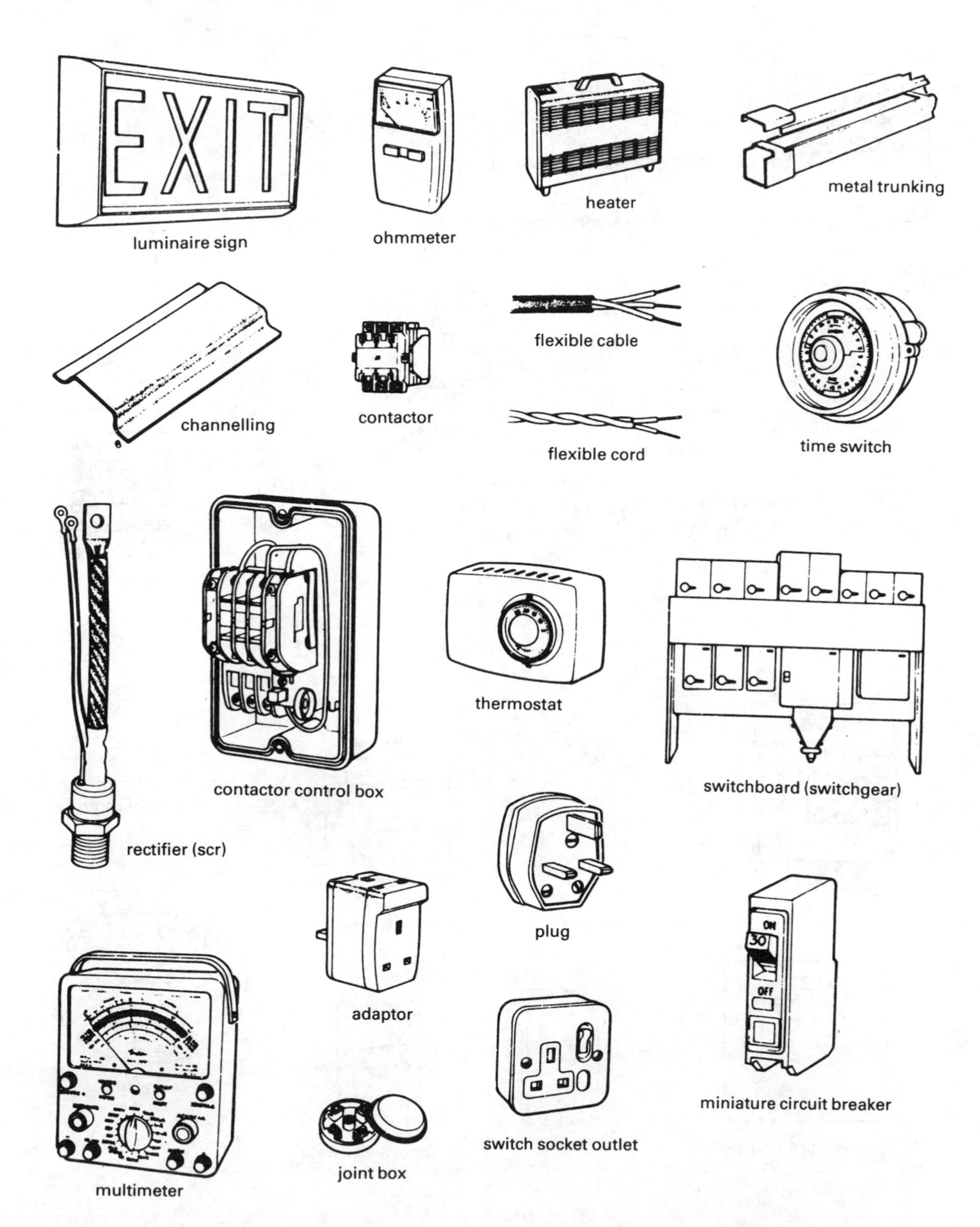

Figure 1.3 *Electrical components*

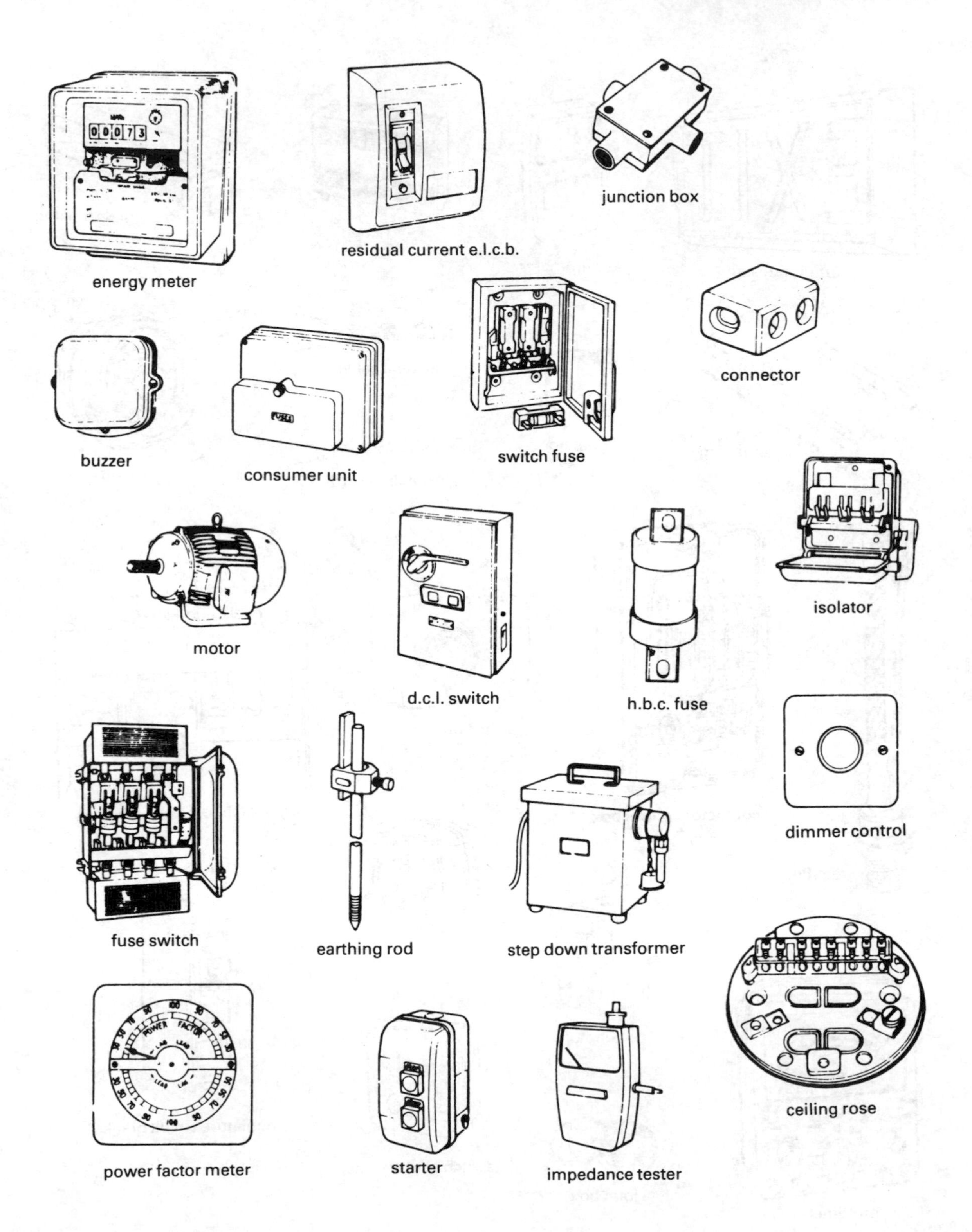
energy meter
residual current e.l.c.b.
junction box
buzzer
consumer unit
switch fuse
connector
motor
d.c.l. switch
h.b.c. fuse
isolator
dimmer control
fuse switch
earthing rod
step down transformer
ceiling rose
power factor meter
starter
impedance tester

Figure 1.3 (cont.)

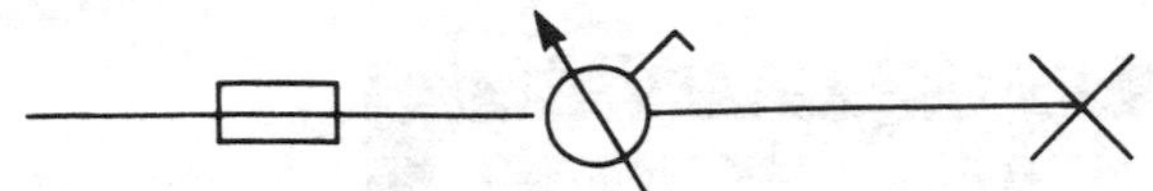

Figure 1.4 *Lighting circuit*

3 Write down meanings for the following:

(a) open circuit
(b) closed circuit
(c) short circuit
(d) conductor polarity
(e) conductor resistance
(f) insulation resistance
(g) fusing factor
(h) space factor
(i) continuity
(j) emergency switching.

4 Explain with the aid of a diagram how a residual current device operates.

5 Figure 1.5 shows the rooms in a three-bedroom house. Insert in each room the appropriate BS 3939 installation graphical symbols for both the lighting and small power circuits. In order to distinguish between the two circuits, use coloured pencils.

6 With reference to the IEE Wiring Regulations, write down the letters and subscripts for denoting the following terms.
For example, design current is abbreviated 'I_b'.

(a) Nominal voltage to earth
(b) Prospective fault current
(c) Current-carrying capacity
(d) Effective operating current (of an excess current device)
(e) Group correction factor
(f) Earth fault loop impedance
(g) Resistance of phase conductor
(h) Cross-sectional area of protective conductor
(i) Nominal current of a circuit protective device
(j) Resistance of protective conductor.

7 With reference to Appendix 4 of the IEE Wiring Regulations, what does the following expression mean?

$I_b \leq I_n \leq I_z$

8 Distinguish between the following terms:

(a) switchfuse and fused switch
(b) junction box and joint box
(c) motor and generator
(d) wattmeter and energy meter
(e) m.c.b. and r.c.d.
(f) overload current and short circuit current
(g) direct contact and indirect contact
(h) TN–S earthing system and TN–C–S earthing system
(i) Class I equipment and Class II equipment
(j) Category 1 circuit and Category 2 circuit.

9 Using BS 3939 installation graphical symbols, show the sequence of control and protection for a domestic consumer's electrical installation. This should be drawn as a line diagram.

10 Draw freehand sketches of the following electrical items:

(a) a cartridge fuse
(b) a 13 A plug
(c) a male brass bush
(d) a junior hacksaw
(e) a distance saddle.

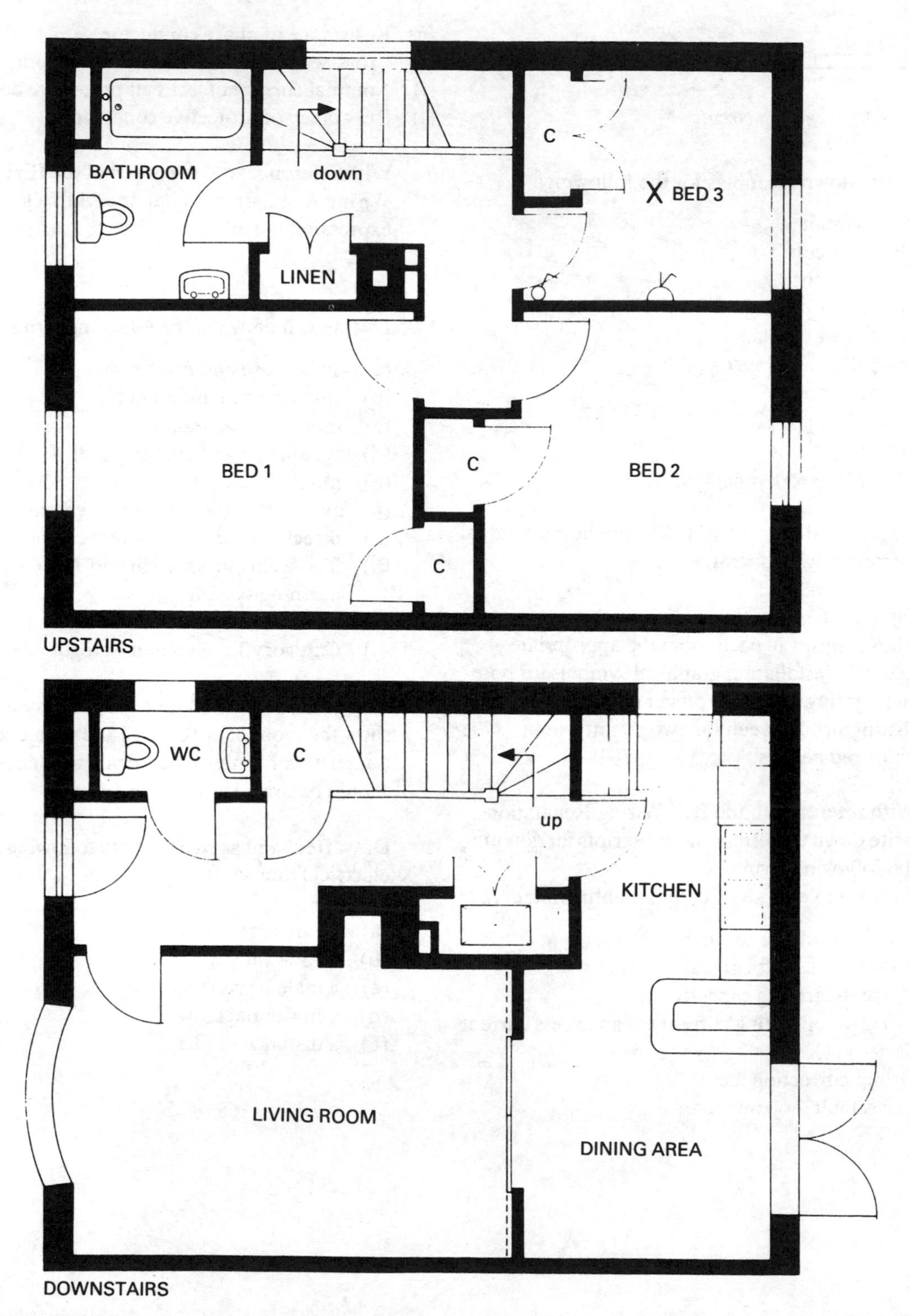

Figure 1.5 *Plan view of a three-bedroomed system*

chapter two

Health and safety

After reading this chapter you should be able to:

1 state the purpose of the following regulations:
   (i) Health and Safety at Work, etc. Act 1974
   (ii) Electricity at Work Regulations 1989
   (iii) IEE Wiring Regulations 1991,

2 state the duties imposed on employers and employees by the above regulations,

3 describe some of the protective measures associated with safe working,

4 know the correct procedures for carrying out first aid and treatment of electric shock,

5 describe types of fire extinguisher, stating their application for different fires,

6 know the procedure for completing a permit-to-work document and an accident form.

Health and Safety at Work, etc. Act 1974 (HSW Act)

The main purpose of the HSW Act is to provide the legislative framework to promote, stimulate and encourage high standards of health, safety and welfare of persons at work, as well as others, against risks to health or safety in connection with the activities of persons at work. The HSW Act has four main parts but it is Part I which concerns the important provisions for people at work. Its objectives set out the 'general purpose' requirements to:

(a) secure the health, safety and welfare of all persons at work,
(b) protect persons other than those at work against risks to their health and safety arising out of work activities,
(c) control the keeping and use of explosive or highly flammable or otherwise dangerous substances,
(d) control the release into the atmosphere of noxious or offensive substances from premises.

The Act established two corporate bodies, namely the Health and Safety Commission (HSC) and Health and Safety Executive (HSE). The HSC is responsible for promoting the general objectives of the Act, encouraging research and training, providing information such as proposals for new regulations and preparing codes of practice. The HSE, on the other hand, acts on behalf of the Commission and has a number of inspectorate staff and supporting staff in the field dealing with factories, agriculture, quarries, mines, railways, nuclear installations, etc. The HSE has the power to enforce statutory requirements on employers so that they provide and maintain a safe and healthy place of work. Inspectors may visit the work place, collect evidence from it, prohibit dangerous working practices, issue improvement notices or prohibition notices which can stop work immediately. The inspectors have both advisory duties and enforcing duties and they can take cases to court. In a Magistrates' Court, fines of up to £2000 can be imposed on offenders who contravene any requirements of the Act, whereas in a Crown Court fines are unlimited and there could even be the possibility of a gaol sentence for up to two years.

In Section 2, every employer has to ensure '*so far as is reasonably practicable*', the health, safety and welfare at work of all his employees in relation to:

(a) the provision and maintenance of plant and systems of work,

(b) arrangements for ensuring safety and absence of risks to health in connection with the use, handling, storage and transport of articles and substances,
(c) the provision of information, instruction, training and supervision,
(d) the maintenance of any place of work under his control in a condition which is safe and without risks to health, as well as the provision and maintenance of means of access to and egress from it that are safe and without such risks, and
(e) the provision and maintenance of a working environment that is safe, without risks to health and adequate as regards facilities and welfare arrangements at work.

Note: The term '*so far as is reasonably practicable*' often appears in statutory instruments and means that the degree of risk has to be balanced against such things as time, cost and the physical difficulty of taking measures to avoid the risk—the greater the degree of risk the less weight that can be given to the cost of measures needed to prevent it.

There are also duties imposed on designers, manufacturers and suppliers as well as importers and hirers of equipment. Employees also have duties under Section 7 of the Act. They have to:

(a) take reasonable care for the health and safety of themselves and of other persons who may be affected by their acts or omissions at work, and
(b) as regards any duty or requirement imposed on their employer or any other person by or under any of the relevant statutory provisions, to co-operate with him so far as is necessary to enable the statutory provision to be performed or complied with.

Section 8 of the Act states:

'No person shall intentionally or recklessly interfere with or misuse anything provided in the interests of health, safety or welfare in pursuance of any of the relevant statutory provisions.'

Under Section 2(3) of the HSW Act, all employers who employ five or more employees have a statutory duty to draw up a written statement of general policy with respect to their health and safety at work. This should:

(a) state the general policy on health and safety,
(b) describe the organisation and arrangements for carrying out the policy,
(c) be brought to the notice of all employees, and
(d) be revised whenever appropriate, and every revision must be brought to employees' attention.

The policy statement has to be signed by the employer who may delegate duties to managers and supervisors but all individuals at every level will have to accept degrees of responsibility for carrying out the policy. What is important is that employees should be able to see from the statement how they fit into the system and what their own duties are and to whom they should go for advice.

Electricity at Work Regulations 1989

This is a statutory instrument made under the HSW Act and came into force on 1 April 1990. It replaced the outdated Electricity (Factories Act) Special Regulations of 1908 and 1944 which covered the generation, transformation, distribution and use of electrical energy in such premises as factories, substations, building operations and works engineering construction. Over the years, it was realised that many other types of premises fell outside its scope such as hospitals, colleges, offices and shops, etc. The new Regulations cover these premises and differ from the old legislation in that they make no distinction between high voltage and low voltage, there are no in-built exemptions and fewer definitions, duties are placed on all those concerned with electrical work, live working is discouraged, there are requirements for adequate supervision, and there is no age limit placed on competent persons.

The main purpose of the new Regulations is to require precautions to be taken against the risk of death or personal injury from electricity in work activities and applies to all work situations and electrical equipment. Like the special regulations, it is supported by a Memorandum of Guidance which is

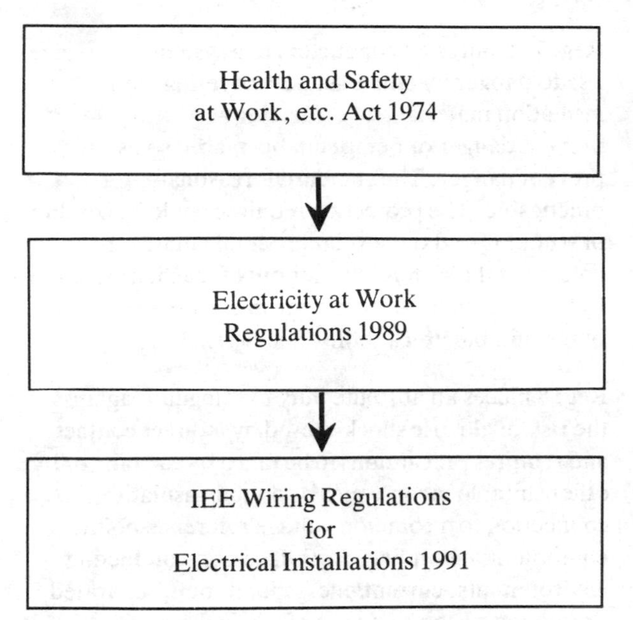

Figure 2.1 *Statutory and non-statutory regulations*

intended to assist '*duty holders*' in meeting the requirements of the Regulations. The purpose of the Memorandum is to amplify the nature of the precautions in general terms so as to help achieve a high standard of electrical safety. Reference is made to the IEE Wiring Regulations and other sources of guidance. Where the qualifying term '*so far as is reasonably practicable*' is absent, the regulation is *absolute* and must be met regardless of cost or any other consideration. A summary of some of the regulations is given below.

Reg. 1 refers to its citation and commencement.

Reg. 2 defines terms frequently used, such as: '*danger*', '*injury*', '*conductor*', '*circuit conductor*', '*electrical equipment*' and '*system*'.

Reg. 3 places levels of duty on employers, the self-employed and employees to comply with the provisions of the Regulations in so far as they relate to matters which are within their control.

Reg. 4 refers to systems, work activities and protective equipment. The words 'construction' and 'maintained' are used with reference to systems where there is a duty to prevent danger. Every work activity, including operation, use and maintenance of a system and work near a system must also be carried out in such a way as not to give rise to danger. The standard is reasonably practicable in 4(1), 4(2) and 4(3) but in 4(4) the regulation is absolute, requiring protective equipment to be suitable for the use for which it is provided and also that it should be suitably maintained and properly used. This Regulation calls for regular inspection and testing for all systems. The Memorandum states that it should form part of a preventative maintenance programme keeping

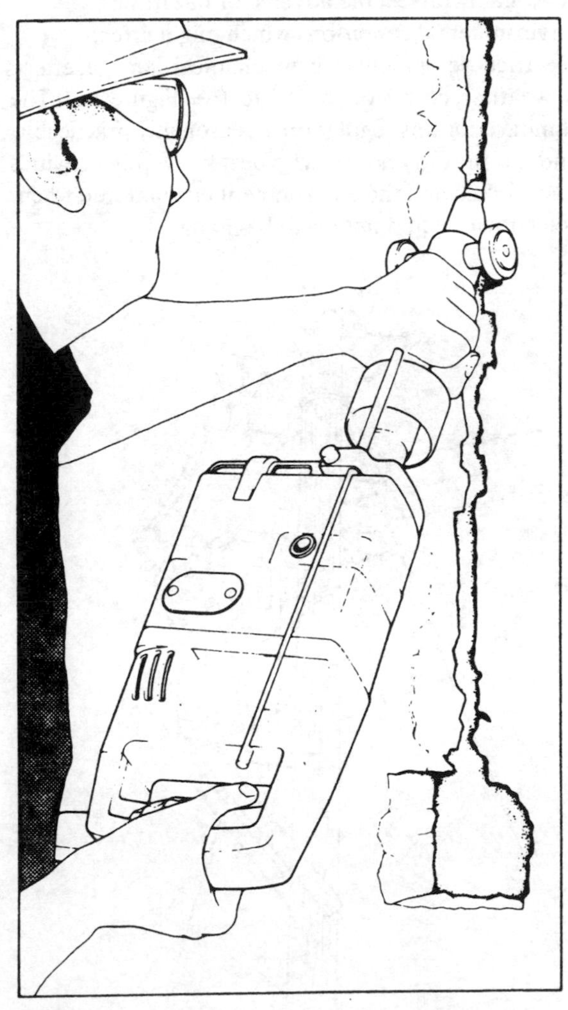

Figure 2.2 *Portable hammer-drill designed for a dusty environment*

records of work and this should include all portable equipment connected to the system.

Reg. 5 is to ensure that the strength and capability of electrical equipment in use are not exceeded and must not give rise to danger. The standard is absolute and the duty holder and system designer will need to ensure that all system components are appropriate to the intended use and rated accordingly. Fault levels should normally be within the equipment's capacity but these can be exceeded provided danger does not arise.

Reg. 6 requires full account to be taken of any reasonably forseeable adverse or hazardous environmental conditions which might affect electrical equipment, e.g. mechanical damage, effects of weather, corrosion, dust, etc. (see Figure 2.2). The standard for this regulation is reasonably practicable and it is the duty holder who best knows the activities undertaken and the environment encountered when selection of equipment is to be made.

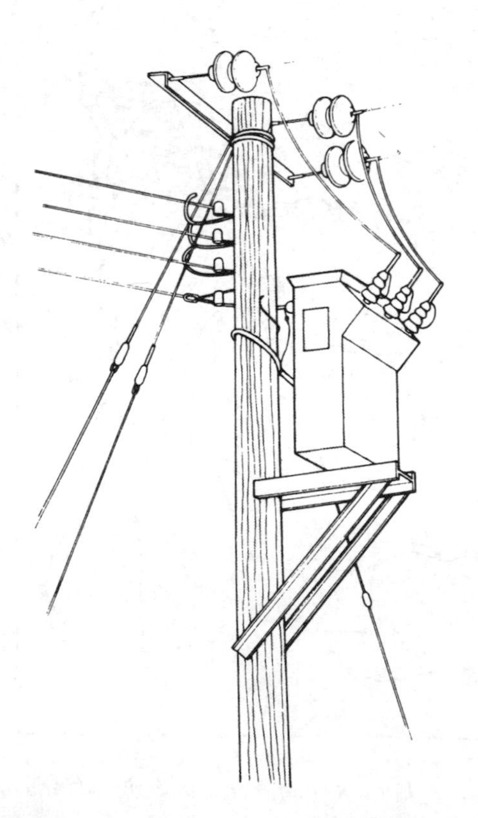

Figure 2.3 *Protection by placing out of reach*

Reg. 7 requires all conductors in a system not to give rise to danger by either suitably covering with insulation material and protecting as necessary to prevent danger or being suitably placed so as to prevent danger. The standard is reasonably practicable. The protection required could be conduit or trunking and the insulation should match the voltage of the system. Placing out of reach, protection by barriers, enclosures or by obstacles are examples of other suitable precautions. See Fig. 2.3.

Reg. 8 places an absolute duty to safeguard against the risk of electric shock caused by indirect contact and requires precautions to be taken by earthing or by other suitable means, such as double insulation, connection to a common voltage reference point, equipotential bonding, earth-free non-conducting environments, current/energy limitation, separated or isolated systems and use of safe voltages.

Reg. 9 seeks to preserve the integrity of referenced conductors such as a PEN conductor or neutral conductor which is normally earthed at the star point of the supply source. The requirement prohibits a fuse or a semiconductor device in the referenced conductor but allows a linked switch or circuit breaker which is designed to break all related phase conductors. The object of this regulation is to prevent the reference conductor from reaching significantly different potentials that would give rise to possible danger.

Reg. 10 is a requirement for all joints and connections in a system to be mechanically and electrically suitable for use. The aim is to prevent danger. All connections in any system conductors, including those to terminals, plugs and sockets and crimped or sweated joints should be suitable for their purpose. Plugs and socket connectors containing a protective conductor should ensure it is made first and broken last and any combinations of metals liable to produce damaging electrolytic action should be avoided. The regulation is absolute and applies irrespective of whether the installation is permanent or temporary.

Reg. 11 recognises that faults and overloads may occur on electrical systems and requires efficient means, suitably located, for protection against excess of current. This is an absolute requirement but compliance with it may be technically impossible or even undesirable since it might create other hazards. In some situations, interruption of the supply by a protective device might cause danger at the point of fault during the finite time taken for operation and in other circumstances a danger might result if a circuit was interrupted and the supply broken, as in the case of lifting electromagnets and secondary windings of current transformers.

Reg. 12 requires suitable means for cutting off the supply and the isolation, i.e. the disconnection and separation (see Figure 2.4) of any electrical equipment. The aim is to ensure that work can be undertaken on an electrical system without danger in compliance with Reg. 13 (i.e. work when equipment is made dead). The regulation recognises the impracticability of switching off or isolation in some instances and requires precautions to be taken in order to prevent danger (see Fig. 2.5).

Reg. 13 requires adequate precautions to be taken on electrical equipment which has been made dead in order to prevent danger while work is carried out on or near the equipment. This regulation is absolute and covers the preferred system of work, i.e. based on working on a de-energised system. The following would form the basis of a typical safe system of work: isolation from all points of the supply; securing each point of isolation; earthing where appropriate; proving dead at point of work; demarcation of the safe zone of work; where necessary, safeguarding from adjacent live conductors; and release for work by the issue of a safety document (see permit-to-work document Figure 2.10).

Reg. 14 imposes an absolute duty such that no person shall be engaged in any work activity on or so near any live conductor where danger may arise unless it can be properly justified. Where live working is properly justified then precautions considered necessary to avoid danger are: (a) trained and competent staff; (b) provision of adequate information; (c) use of suitable insulated tools, equipment and protective clothing; (d) use of suitable insulated barriers or screens; (e) use of suitable instruments and test probes and consideration of the need for accompaniment; and (f) control of working area. Here, duties require the employer to provide free any protective clothing or equipment; report certain injuries, diseases and dangerous occurrences; provide adequate first-aid facilities; and also the taking of precautions against fire.

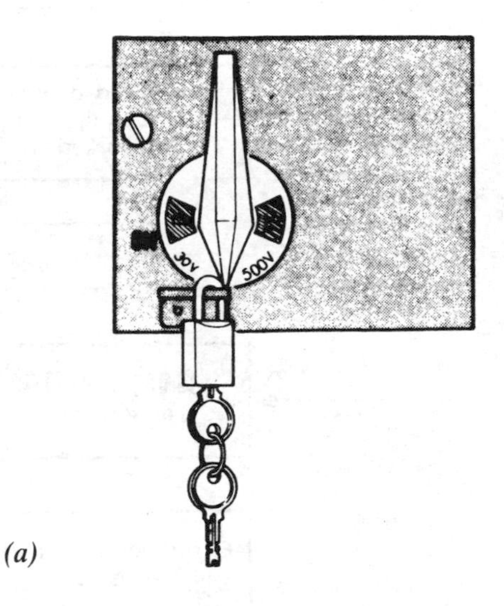

(a)

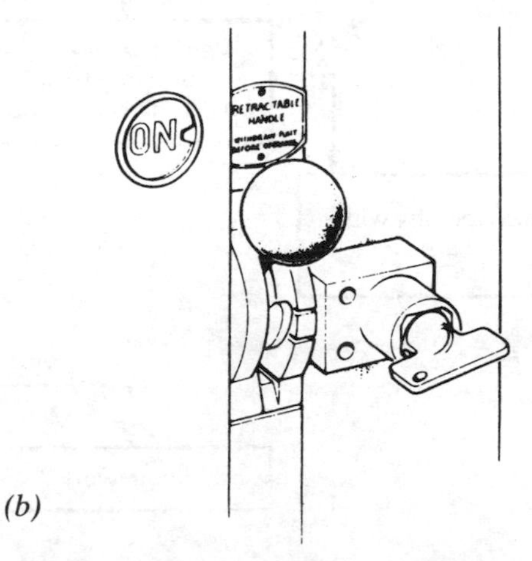

(b)

Figure 2.4 *Supply isolation: (a) locking off an isolator; (b) locking off switchgear using a Castell interlock*

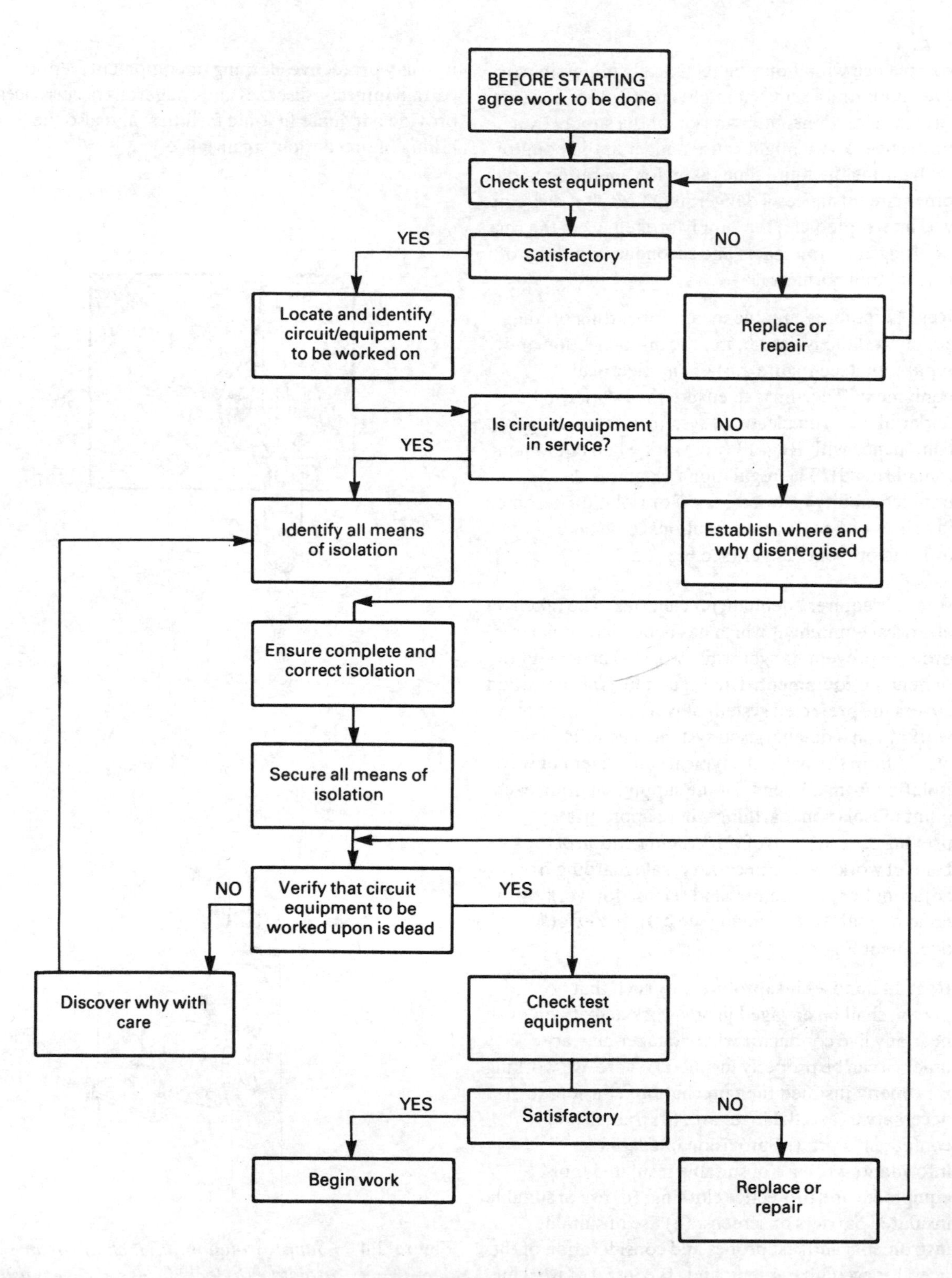

Figure 2.5 *Procedure for isolation of supply before work commences*

Reg. 15 requires the provision of adequate working space, adequate means of access, and adequate lighting for all electrical equipment on which or near which work is being done to avoid danger. The system designer must specify working space requirements (a) for the operation of equipment and replacement of fuses, etc. and (b) the carrying out of work on equipment. Appendix 3 of the Memorandum gives guidance on the requirements for access clearances for live conductors depending on the system voltage and there is an HSE Guidance booklet HS(G) 38 which provides recommended minimum levels of illuminance for different work activities.

Reg. 16 refers to persons to be competent to prevent danger and injury. It requires no person to be engaged in any work activity unless they have the technical knowledge or experience or are under such degree of supervision as may be appropriate having regard to the nature of the work. The regulation is absolute and has considerable implications concerning the provision of sufficient appropriate training. This is particularly relevant in the case of those staff expected to carry out work on live conductors in accordance with requirements imposed by Reg. 14 or to undertake work or operate high-voltage systems.

IEE Wiring Regulations 1991

This is a very important code of practice for the designer, installer and tester of an electrical installation and is known to the industry as the 16th Edition. Its purpose is to protect persons, property and livestock against electric shock, fire, burns and injury from mechanical movement of electrically actuated equipment.

The 16th Edition makes considerable reference to British Standards as well as related statutory regulations.

The 15th Edition of the IEE Wiring Regulations had 17 appendices and many of these have now been moved to guidance notes on systems operating at 240/415 V and 100 A maximum loading. These guidance notes will cover: (a) the service position, (b) final circuits, (c) special locations and (d) inspection and testing.

Individual regulations in the 16th Edition are identified by three-part numbering (e.g. voltage drop in consumers' installations is covered by Regulations 525–01–01 and 525–01–02). The first digit (5) is the part number and inclusion of the next digit (2) creates the chapter number (52). Inclusion of a third digit (5) creates the section number (525). The second part number after the hyphen with inclusion of the fourth and fifth digits (01) cover the group regulation number (525–01). The two specific regulations mentioned then follow the second hyphen (see Figure 2.6). Some general points about the 16th Edition are as follows.

Part 1 is concerned with scope, object and fundamental requirements for safety. Electricity supply authority's work, work in mines and quarries are some installations excluded from the scope. The Regulations are designed to protect persons, property and livestock from electric shock, fire, burns and injury. Failure to meet the requirements of Chapter 13 may not guarantee a supply connection to a consumer's premises.

Part 2 contains numerous definitions such as 'earthed equipotential bonding' and 'prospective fault

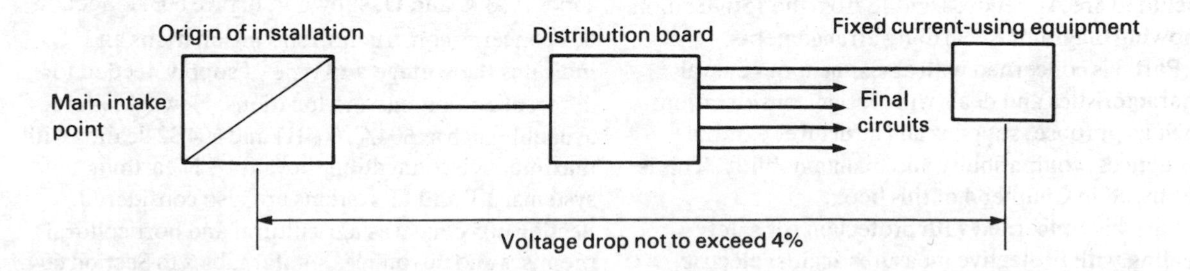

Figure 2.6 *Satisfying Regulation 525–01–02*

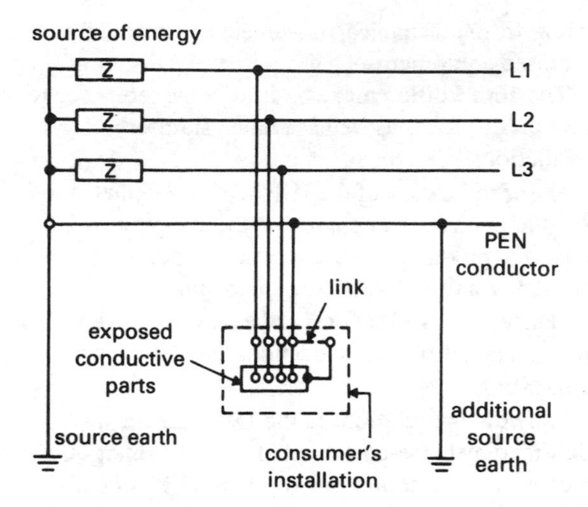

(a) TN-C-S system

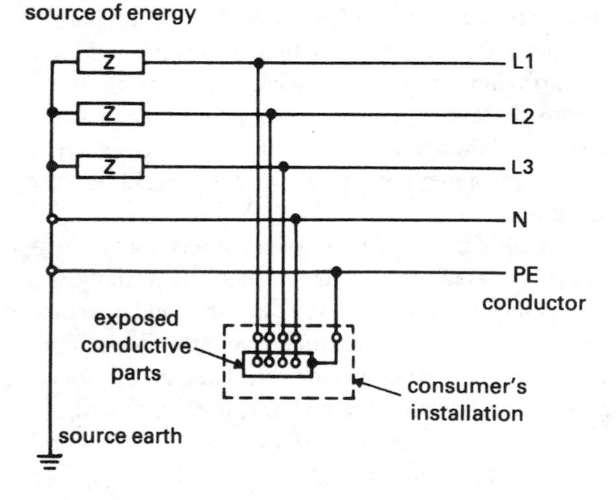

(b) TN-S system

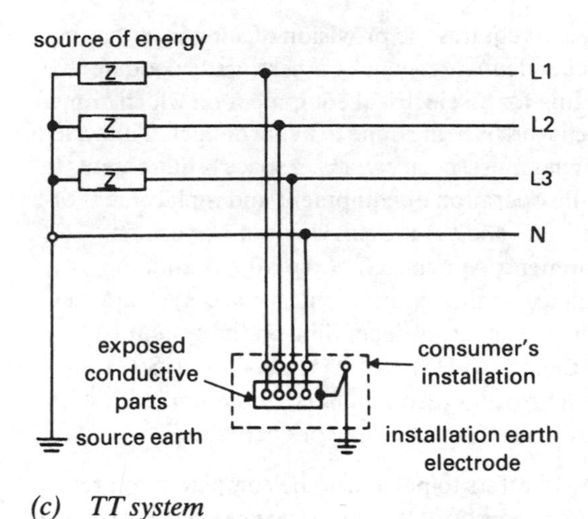

(c) TT system

Figure 2.7 *Typical earthing systems in the UK*

current', which now includes faults to earth. Also included are Appendix 3 and 13 from the 15th Edition showing diagrams of earthing arrangements.

Part 3 is concerned with assessment of general characteristics and deals with design considerations such as purpose, supplies and structure, external influences, compatibility and maintainability. This is discussed in Chapter 4 of this book.

Part 4 is concerned with protection for safety dealing with protective measures against electric shock, thermal effects, overcurrent, undervoltage and isolation and switching.

Part 5 is concerned with selection and erection of equipment, covering common rules, selection of wiring systems, switchgear, earthing arrangements, other equipment and supplies for safety services.

Part 6 is concerned with special installations or locations. There are currently nine sections in use; 601 to 608 and 611. Sections have no specific chapters but all provide a scope and all make clear the protection for safety, either against electric shock or against overcurrent or both. Section 601 concerns locations containing a bath tub or shower basin and introduces new regulations concerning wiring systems, devices for isolation and switching, transformers, plugs and socket outlets, luminaires and fixed heating equipment. Section 602 concerns swimming pools and introduces the concept of zoning, shown in Figures 602A and 602B of the Regulations, for assessment of general characteristics. Section 603 concerns hot air saunas and states requirements on four temperature zones A,B,C and D, shown in Figure 603A. Section 604 concerns construction site installations and indicates the voltage and type of supply needed for different equipment and locations. New tables are available such as 604A, 604B1 and 604B2 dealing with maximum disconnection times for TN earthing systems. TT and IT systems are also considered. Section 605 concerns agricultural and horticultural premises and introduces similar tables to Section 604. It also introduces requirements for protection against fire and harmful thermal effects as well as electric fence controllers (see Fig. 2.8). Section 606 is concerned with conductive locations where freedom of personal movement is restricted, mainly by the dimensions of the location. Requirements for protection against both direct and indirect contact are fully stated.

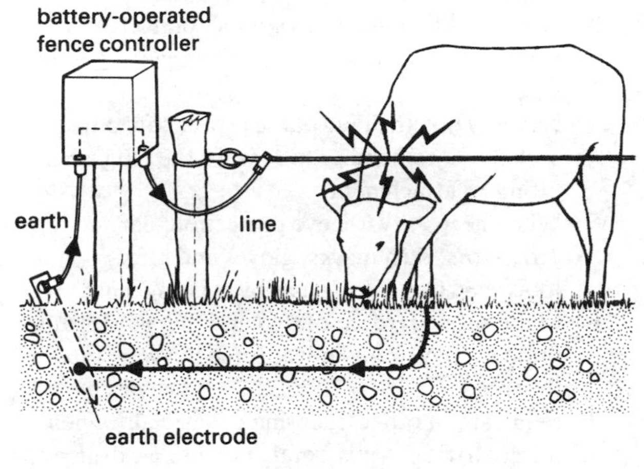

Figure 2.8 *Application of electrical fence*

Section 607 concerns equipment having high earth leakage currents. Such equipment shall not be connected to an IT earthing system. Where low noise earthing is specified, reference is to be made to Sections 545 and 546. Section 608 concerns caravans and motor caravans (division 1) and caravan parks (division 2). It revises the old Section 471 and Appendix 11 and introduces new requirements and a notice of supply instruction and periodic inspection. A new Table 60A on flexible cord sizes for caravan connections is provided. Section 611 concerns highway power supplies and street furniture and covers the needs of these and similar installations, whether or not they are situated on or near a public highway.

Part 7 is concerned with inspection and testing, covering initial verification, periodic inspection and testing, and certification and reporting.

Appendix 1 covers relevant up-to-date British Standards to which reference is made; Appendix 2 concerns statutory regulations and associated memoranda relating to listed installations; Appendix 3 concerns information on time/current characteristics of overcurrent protective devices and provides new tables for each device. Appendix 4 concerns current-carrying capacity and voltage drop for cables and flexible cords and has been revised in numerous places with different table headings and table amendments. Appendix 5 is concerned with the classification of external influences developed for IEC Publication 364 and completely revises Appendix 6 of the 15th Edition. Appendix 6 is concerned with forms of completion and inspection certificates.

Electrical safety

During the period between January 1981 and March 1988, 109 fatal accidents to employees were reported to HM Factory Inspectorate. Forty-four (40%) of these accidents happened to electrical people. The main cause (61%) of these accidents was the failure to isolate the electrical system. Sixty-nine (63%) of the fatal accidents were attributed to management for not introducing a safe system of work. This is a requirement by law. Isolation means disconnection and separation of electrical equipment from all sources of electrical energy—de-energised plant must not inadvertently become live again while being worked upon (see Reg. 12, EWR 1989).

The following are some examples of the electrical accidents that happened in the survey.

A groundsman received a fatal electric shock while filling a kettle with water. The kettle was plugged in and switched on at the time of the accident. The kettle was five years old but had not been used for about 18 months since an electric water heater had been provided in the mess room. However, the water heater had failed and while repairs were awaited the kettle was taken back into use. Examination of its lead after the accident showed that the earth wire was not connected to the terminal of the plug. The insulation had melted at the point where the live and earth wires crossed.

An electrician employed by a contractor was electrocuted while installing security cameras in a car showroom. At the time, he was working from a metal step ladder running cables through the false ceiling. It was thought that he inadvertently touched the live terminals of a three-pin plug lying unprotected on top of a display lighting gondola near his work. He was thrown from the ladder by the shock and died in hospital. After investigation it was found that the live pin in the plug connector had been wrongly wired.

A security guard was found electrocuted while attempting to rectify a fault on a 1 kW electric bar-type fire. The fault had developed on a previouis shift and he was found gripping the live bar with one hand while the other was holding the earthed metalwork of the fire.

A deputy plant electrical engineer sustained fatal burns to his head, arms and body while carrying out maintenance on a live 415 V switchboard. With another employee he was attempting to change a contact which had overheated. Using an uninsulated spanner he touched the live busbar connections causing a short circuit to earth. He received severe burns as he was leaning into the switchboard at the time. A permit-of-work system had been drawn up which required the engineer to isolate the switchboard thereby shutting down the plant. The switchboard had not been isolated. The engineer was the most senior electrically qualified staff member at the plant.

Power tool safety hints

The market is flooded with different types of power tools and a number of these are provided with additional attachments to extend the range and application. A large number of accidents occur because of misuse and by failing to have portable tools regularly serviced by specialists in the field. General safety measures ought to reflect the following points.

1. Always read the manufacturer's literature first.
2. Check the power tool for correct voltage, frequency, fuse protection and correct speed of operation.
3. Avoid holding the tool's contact button when carrying it, especially if it is still connected to the supply.
4. Never 'yank' the plug of the power tool out of its socket and remember to remove the plug when fitting an attachment.
5. When necessary use eye protection, ear protectors, dust masks, gloves and safety footwear. Cover long hair by wearing a hat.
6. Make sure guards are fitted and that guides are fitted correctly, and never force the workrate of the tool.
7. Be aware of other tradesmen, especially when using cartridge tools, rotary percussion-drills and rotary hammer-drills.
8. Keep fingers clear when drilling holes and tighten drill chucks with the proper key. Keep the chucks clean and in good order.
9. Protect the cable leads from mechanical damage.
10. Avoid over-reaching when using a power tool up a ladder. It is better to work off a platform, staging or scaffolding, with both feet on a firm foundation. Figure 2.9 illustrates care in using a portable tool.

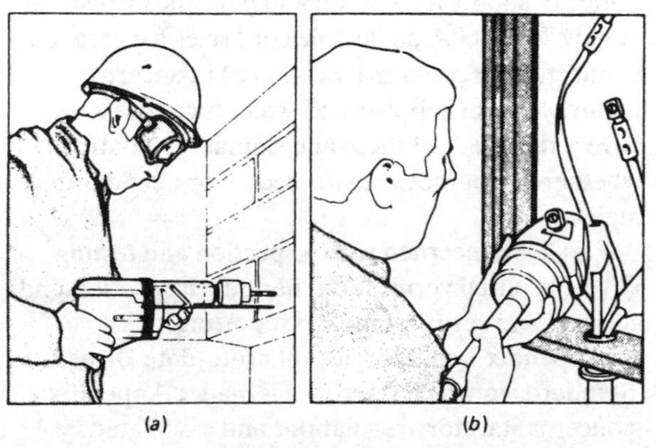

Figure 2.9 *Taking care using portable tools: (a) wearing protective gear; and (b) showing a safe termination method*

11 Electrical work should always be carried out by competent staff.
12 Electrical equipment should be regularly inspected and tested.
13 Always seek permission or a permit-to-work document when servicing live switchgear.
14 Ensure that warning notices are displayed and main switches are locked in the off position.
15 Test circuits to see if they are dead and remember to test the tester first. It is not enough to rely on verbal instructions or even fuseboard markings.
16 Where live work is unavoidable, wear rubber-soled shoes or stand on a dry rubber mat. Check that the tools are well insulated and sound in construction.
17 Do not attempt to hold exposed metal parts that are earthed and exercise extra care when disconnecting live conductors from one terminal post—remember that the neutral is a live conductor.
18 Faulty equipment should be disconnected, labelled and sent for repair, and portable equipment should be serviced regularly.
19 When terminating flexible cables, make sure that the polarity of connections is correct.
20 Never hang portable tools by their flexible leads and avoid 'knotting' or 'kinking' the leads when coiled up. Use appropriate ties and proper cable reels for extension leads.

Hand tools

Today, an electrician will have numerous hand tools and they will frequently require checking in order to maintain a high standard of good workmanship. The majority of hand tools are designed for one particular purpose and, because of this, every effort ought to be made to select the correct one for the job. Some useful tips are the following.

1 When using a spanner, make sure you select the correct size for the job and try not to push it away from your body. Keep it free from oil or grease whenever possible and do not use it as a lever.
2 When using a hammer, avoid using the shaft as a lever or a striker. Redress the handle if it splinters and make sure the head is securely attached to its shaft. Select the correct one and use eye protection if necessary.
3 When using pliers, select the right size and type for the job and avoid finger traps when a sudden release occurs. Ensure that the cutting edges remain sharp or replace the defective tool. Check insulated pliers regularly.
4 When using screwdrivers, select the correct size for the job. The tip should fit the screw slot and it should not be wider than the screwhead. Exercise care and avoid oil or grease on the handle as well as an over-sharp, irregular and deformed blade. Do not strike the screwdriver with a hammer or use it as a lever. If grinding the blade, use eye protection.
5 When using a lever such as a crowbar or even a jack, avoid traps to the hands and feet, especially when rollers are used. Be aware of the sudden collapse of any part under stress and, when using a jack, it is better to supplement it with suitable blocks where necessary. Props should be sensibly placed and strong enough to support the load. Also, ensure that lifting tackle is efficient and be aware of objects that are slippery. It is advisable to wear safety shoes.
6 When using files and chisels, again select the correct ones for the job. Avoid using a file without its handle and do not use it as a lever or to stir paint. Avoid the formation of a mushroom head on a chisel and examine the handle frequently. Always chisel away from your body and keep the tool sharp.
7 When using saws and other tools with blades, make regular checks on their sharpness and their teeth. Always securely fasten blades to their attachments and do not use them outside their frames. Keep them free of rust and always return them to their special boxes.

General safety

1 If you are called upon to lift and handle material, size up the job first. For example, will it require

BS 5405 : 1976

MODEL FORM OF PERMIT-TO-WORK – FRONT

THE ELECTRICITY BOARD

DISTRIBUTION PERMIT-TO-WORK

1. ISSUE No.

To

I hereby declare that it is safe to work on the following H.V. Apparatus, which is dead, isolated from all live conductors and is connected to earth: –

....

....

....

ALL OTHER APPARATUS IS DANGEROUS

Points at which system is isolated

....

....

....

Caution Notices posted at

....

....

....

The apparatus is efficiently connected to earth at the following points

....

....

....

Other Precautions

....

....

....

The following work is to be carried out

....

....

....

Signed ....
being a Senior Authorised Person.

Time............................Date............................

MODEL FORM OF PERMIT-TO-WORK – BACK

2. RECEIPT

I hereby declare that I accept responsibility for carrying out the work on the apparatus detailed on this Permit-to-Work and that no attempt will be made by me, or by the men under my control, to carry out work on any other apparatus.

Signed..

Time............................Date............................

Note: After signature for the work to proceed this Receipt must be signed by and The Permit-to-Work be retained by the person in charge of the work until the work is suspended or completed and the Clearance section has been signed.

3. CLEARANCE

I hereby declare that the work for which this Permit-to-Work was issued is now *suspended/completed, and that all men under my charge have been withdrawn and warned that it is no longer safe to work on the apparatus specified on this Permit-to-Work, and that gear, tools and additional earthing connections are all clear.

Signed..

Time............................Date............................

** Delete word not applicable.*

4. CANCELLATION

This Permit-to-Work is hereby cancelled.

Signed..
being a Senior Authorised Person possessing authority to cancel a Permit-to-Work.

Figure 2.10 *Permit-to-work document*

some assistance? Is the floor area slippery? What obstacles are in the way? Are there any sharp edges likely to cause injury? Will protective clothing be required?

2 Lifting a fairly heavy object from the floor requires a certain amount of skill and correct lifting posture. The procedure is to bend your knees and crouch close to the load. Obtain a firm grip with elbows tucked in and then keep your head and back as straight as possible. Lift the load by straightening your knees and use your body weight to assist in the movement (see Figure 2.11).

3 Avoid wearing finger rings or a wrist-watch, or any loose clothing such as a tie or scarf, as these items could act as a trap between you and the load.

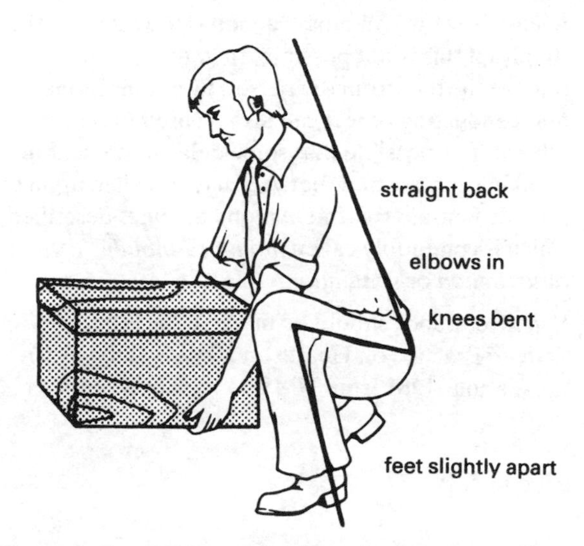

Figure 2.11 *Lifting a load from the floor correctly*

4 If you are going to handle greasy or oily materials, change into protective clothing and use barrier cream on your hands.

5 Where lifting tackle is not available, seek assistance from fellow workmates. Try to carry awkward loads evenly amongst you, being extra careful to avoid hand and foot traps. It is much safer to put heavy loads down on solid supports.

6 When carrying a ladder single-handed, be particularly careful when passing through doorways and approaching entrances. Keep the leading end up to avoid risk of striking someone in front of you. Be extra careful when handling a metal ladder near overhead power cables. You should always try and select a ladder of the correct length for the job. Where a ladder rests against a landing platform, set it out using the 4:1 rule and allow it to extend at least 1 m for stepping on-off purposes. A ladder over 7.6 m in length requires lashing to a support to prevent it slipping sideways (see Figure 2.12).

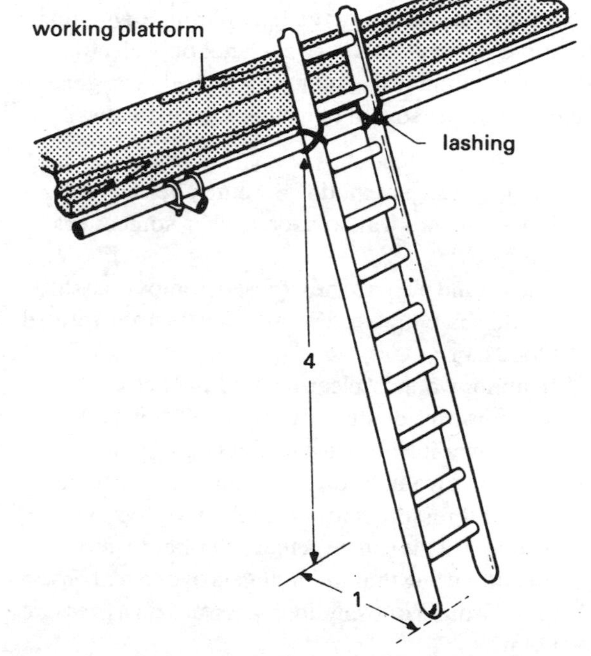

Figure 2.12 *Sound placing and support of a ladder*

First-aid notes

In every factory there must be a first-aid box or cupboard of the prescribed standard, and it should contain nothing except first-aid requisites. Where a large number of people are employed, there should be more than one first-aid box and a responsible person should be placed in charge. Such a person should be trained in first-aid treatment and is required to have a refresher course every three years and have his/her certificate renewed. The responsible person must always be readily available during working hours and a notice must be fixed in every workroom stating the name of the person in charge of the box or cupboard provided.

Where an employee is called upon to play the role of first-aider, his/her actions may be vital in saving life. It is of course appreciated that the first-aider is limited to the assistance rendered at the time of the emergency with such material as may be available, although he/she should stand by after making a report to the doctor if required.

In Regulation 12–1 of the IEE Wiring Regulations, reference is made to safety from not only electric shock, fire and burns, but injury as well. The general rules for the treatment of wounds are as follows.

1 Sit or lie the patient down, remembering that blood escapes with less force when someone is lying down.
2 The wound needs to be exposed, removing as little clothing as possible. Try not to disturb any formed blood clot.
3 In minor wounds, clean around and across the wound with a diluted antiseptic solution and then cover it with a sterile dressing.
4 In serious wounds, do not attempt to wash the wound. Pressure must be applied to stop the bleeding. This can be achieved either by direct pressure using thumbs or fingers over a pad on top of the wound or by indirect pressure on a pressure point.
5 Apply a dressing and then a bandage to the wound.
6 If serious, rest the injured part. If, however, further bleeding occurs, do not remove the dressing but apply an extra bandage.

When a person suffers from a burn, if it is a minor one it need only be placed under a slow-running cold water tap for several minutes. The burnt area should then be covered with a clean dressing. It is advisable to remove any constrictions such as rings and bracelets before the area begins to swell. If the person is badly burnt he should be made to lie down and keep still and wrapped in a clean sheet. The person should be kept warm with blankets if necessary and taken immediately to hospital. In cases of chemical burns, one should flood the area with large amounts of running water. It is important to avoid contaminating yourself. In most cases, treat the affected areas as a wound and take the person to hospital.

Figure 2.13 illustrates the general procedure in the treatment of electric shock. It should be pointed out that action is *urgently required* in these matters and care must be taken not to electrocute yourself. It is rare to find two situations alike and one can only generalise about possible actions taken. If, for example, a person is holding a power tool and cannot release it, your immediate action is to switch off the supply. If this is not possible, then the next step is to remove the tool from the person by pulling its lead. A non-conducting object can also be used to secure release. It is possible that some delay may occur in breaking contact and that you may be called upon to provide first-aid treatment along the lines described, which is commonly called *mouth-to-mouth resuscitation* or *ventilation.*

Note: Reference should be made to the Approved Code of Practice on Health and Safety (First-Aid) Regulations 1981 from HMSO.

Fire safety

The Factory Act 1961 requires every factory to be provided with a means for fighting fires, which shall be appropriately maintained and placed so as to be readily available for use.

For fire-fighting in rooms where there are no exceptional risks to life, portable fire extinguishers are normally sufficient, provided of course the right type is used. Advice on choice of correct extinguisher is provided by the local Fire Brigade. In practice, the extinguishing agent may take several forms:

(a) *Water*—applied as a jet or spray to ordinary combustible materials such as wood. The colour

ELECTRIC SHOCK

ACT AT ONCE—DELAY IS FATAL

1. make sure it is safe to approach

If the casualty is not clear of the source of the shock break the contact by switching off the current, removing the plug or wrenching the cable free. If this is not possible, stand on dry insulating material such as newspaper or a rubber mat and try to push or pull the casualty clear of the contact using a brush or stool. Do not touch the casualty with bare hands but if nothing else is available then grip any loose dry clothing (not under the arms) and pull the casualty away from the source.

2. check for signs of breathing

Place your ear above the casualty's mouth and look along the chest and abdomen for signs of movement.

if the casualty is breathing

Place casualty in the recovery position and call medical aid.

if NOT breathing

Get someone to call medical aid while you begin mouth to mouth ventilation.

Recovery position

3. mouth to mouth ventilation— speed is essential

(● manual method given below)

1. Check airway is not blocked. Turn head to one side and remove obstructions around the face and neck, and any debris in mouth.

2. Open the airway by placing one hand under the neck, the other on the forehead and tilting the head back. Transfer your hand from the neck to the chin and push upwards. Breathing may start. If so place in recovery position.
3. If the casualty is still not breathing. Take a deep breath, pinch the casualty's nostrils together with your fingers and seal your lips around the mouth. Blow air into the lungs and watch the chest rise until it has fully expanded.
4. Remove mouth well away and breathe out excess air. Watch chest fall.

5. Repeat first four inflations as quickly as possible without waiting for lungs to deflate completely. Check the pulse at the neck (carotid) to ensure the heart is beating.

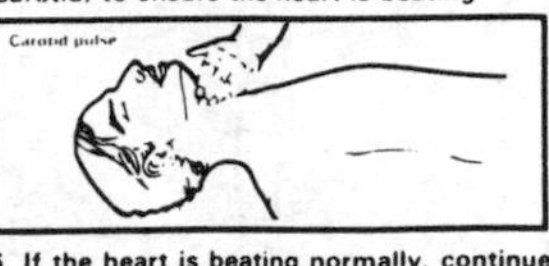

6. If the heart is beating normally, continue inflations at natural breathing (16-18 per min.) rate until normal breathing is restored. Place casualty in the recovery position.

4. if AFTER FOUR VENTILATIONS the casualty does not respond, start external chest compression

Kneel alongside the casualty facing the chest. Place the heel of your hand on the centre of the lower half of the breastbone. Cover this with the heel of your other hand and lock the fingers together. With arms straight, press vertically down on the lower half of the breastbone to depress chest about 4cm (1½ in). Do this 15 times at the rate of 80 per minute.
Reopen the airway and give two ventilations. Repeat the 15 compressions and two ventilations for one minute and check carotid pulse. If present, continue with inflations until natural breathing is restored, then place the casualty in recovery position. If the pulse is absent, repeat cycles, checking for heart-beat every three minutes or 12 cycles.

● manual artificial ventilation method

If mouth to mouth ventilation is not possible (ie if there are severe facial injuries) the casualty should be turned face down with the hands placed under the head. Turn the head sideways and ensure that the airway is open. Kneel in front of the casualty's elbow. Put your hands on the shoulder blades ensuring that the thumbs are either side of the spine. With elbows straight rock forwards until your arms are vertical and leave for two seconds. Rock backwards allowing your arms to slide back and grab the casualty's arms just above the elbow. Raise the arms and pull until tension is felt and leave for three seconds. Be careful not to overstretch. Repeat 12 times per minute. After four cycles check the pulse. If present continue but if not turn the casualty and perform external chest compression.

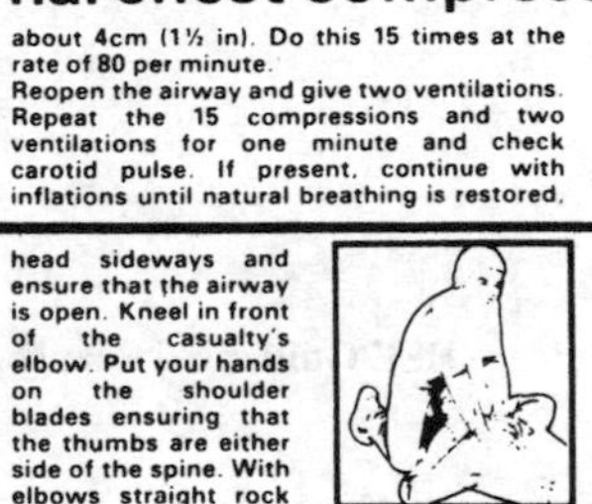

doctor:

phone:

ambulance:

phone:

hospital:

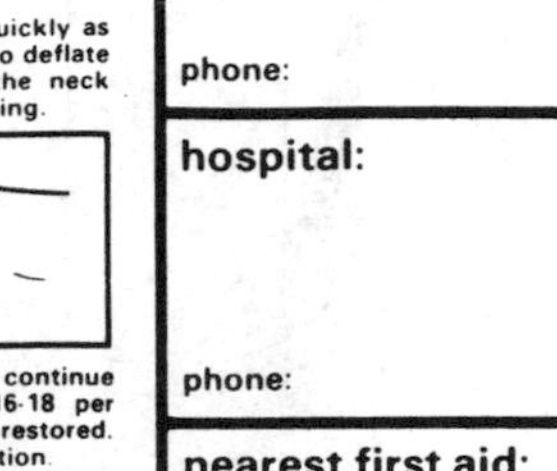

phone:

nearest first aid:

phone:

Prepared with the assistance of the British Red Cross Society.

These instructions also apply to: DROWNING HEART ATTACK SUFFOCATION

Figure 2.13 *Recommended procedures in the event of an electric shock*

of this extinguisher is red and it should not be used on electrical fires.

(b) *Foam*—applied as a jet, which smothers fires caused by oil, petrol or paint. This extinguisher is coloured white or cream and should not be used on electrical fires.

(c) *Vaporising liquid*—applied under pressure to produce a heavy vapour to smother fires caused by flammable liquids or electrical apparatus. Halon (BCF) is the most common extinguisher and is coloured green. It should not be used in confined spaces because it gives off toxic fumes.

(d) *Carbon dioxide*—applied as a jet of gas, which dilutes the oxygen content of the surrounding air and smothers the fire. This extinguisher is coloured black and is used on electrical fires and small flammable-liquid fires.

(e) *Dry powder*—applied as a jet to fires created by flammable liquids as well as metals, plastics and electrical equipment. These extinguishers are coloured blue and their function is to smother the fire.

If an electrical fire does occur, it is important to switch off the supply immediately. *Do not* fight the fire if it becomes too dangerous and if your escape route is threatened. If you have to withdraw, close windows and doors behind you whenever possible.

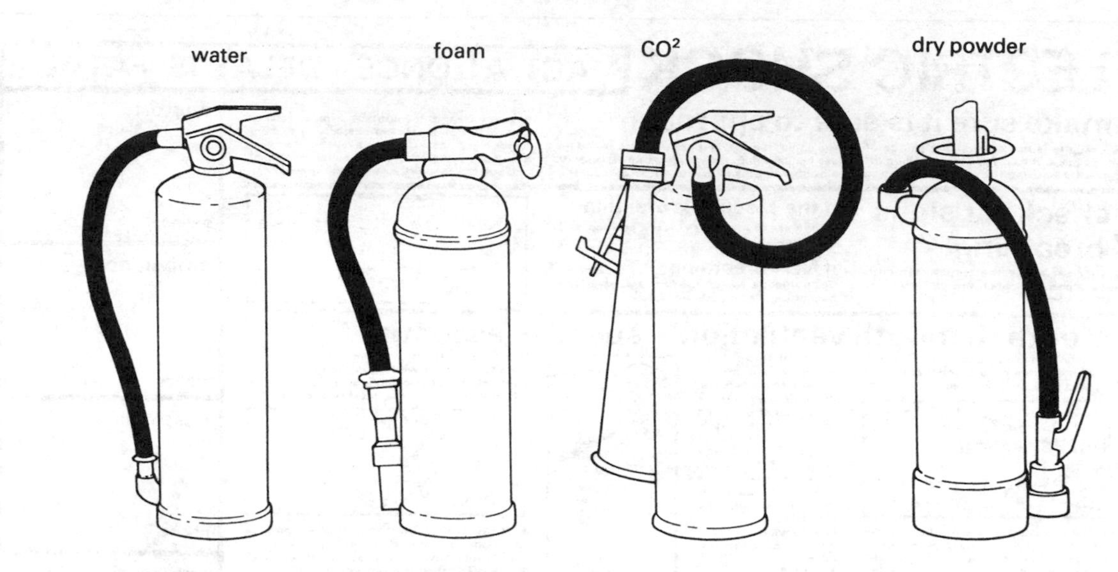

Figure 2.14 *Types of fire extinguisher*

Accident reporting

All accidents that cause death or major personal injury (such as a fractured skull) in a workplace, either to an employee, self-employed person or even to a member of the public, must be reported to the local enforcing authority. Any notifiable dangerous occurrence such as an explosion or the collapse of scaffolding that is more than 5 m high must be reported. An electrical overcurrent accompanied by fire or explosion, resulting in a stoppage of plant for more than 24 h and which may have led to injury or death, must also be reported. The Health and Safety Executive has produced a revised form for the reporting of injuries or dangerous occurrences (*Form F2508*). It is stated that the onus for writing the report is on the employer of the injured person; a self-employed person; someone in control of the premises where work is carried out; or someone who provides training for employment. The main aspects of this form are shown in Figure 2.15 and it is important that the guidance notes be carefully read before it is actually completed.

HSE Guidance Notes (F2508)

1 General

Please use this form for making reports to the enforcing authority about events covered by Regulation 3 or 6 of the Reporting of Injuries, Diseases and Dangerous Occurrences Regulations 1985.

If you do not know who the appropriate enforcing authority is, then send the form to the nearest office of the Health and Safety Executive (HSE). They will pass it on if necessary.

If you are unsure of the address of the nearest HSE office and it is not listed in the local telephone directory, you may find out by telephoning the HSE enquiry point on 071 221 0870.

2 Subject of report

The tick boxes in Section A cover the different kinds of event that must be reported under Regulations 3 and 6.

Box 1

Tick this box if someone has died as a result of an accident arising out of or in connection with work.

Health and Safety Executive
Health and Safety at Work etc Act 1974
Reporting of Injuries, Diseases and Dangerous Occurrences Regulations 1985

Spaces below are for office use only

Report of an injury or dangerous occurrence

- Full notes to help you complete this form are attached.
- This form is to be used to make a report to the enforcing authority under the requirements of Regulations 3 or 6.
- Completing and signing this form does not constitute an admission of liability of any kind, either by the person making the report or any other person.
- If more than one person was injured as a result of an accident, please complete a separate form for each person.

A Subject of report *(tick appropriate box or boxes) – see note 2*

Fatality 1 | Specified major injury or condition 2 | "Over three day" injury 3 | Dangerous occurrence 4 | Flammable gas incident (fatality or major injury or condition) 5 | Dangerous gas fitting 6

B Person or organisation making report (ie person obliged to report under the Regulations) – *see note 3*

Name and address –

Post code –

Name and telephone no. of person to contact –

Nature of trade, business or undertaking –

If in construction industry, state the total number of your employees –

and indicate the role of your company on site *(tick box)* –

Main site contractor 7 | Sub contractor 8 | Other 9

If in farming, are you reporting an injury to a member of your family? *(tick box)* Yes No

C Date, time and place of accident, dangerous occurrence or flammable gas incident – *see note 4*

Date 19 (*day month year*) Time –

Give the name and address if different from above –

Where on the premises or site – and Normal activity carried on there

ENV

Complete the following sections D, E, F & H if you have ticked boxes, 1, 2, 3 or 5 in Section A. Otherwise go straight to Sections G and H.

D The injured person – *see note 5*

Full name and address –

Age Sex (M or F) Status *(tick box)* – Employee 10 | Self employed 11 | Trainee (YTS) 12 | Trainee (other) 13 | Any other person 14

Trade, occupation or job title –

Nature of injury or condition and the part of the body affected –

F2508 (rev 1/86) *continued overleaf*

Figure 2.15 *Form for reporting an injury or dangerous occurrence*

E Kind of accident - *see note 6*

Indicate what kind of accident led to the injury or condition (*tick one box*) –

Contact with moving machinery or material being machined ☐ 1	Injured whilst handling lifting or carrying ☐ 5	Trapped by something collapsing or overturning ☐ 8	Exposure to an explosion ☐ 12
Struck by moving, including flying or falling, object. ☐ 2	Slip, trip or fall on same level ☐ 6	Drowning or asphyxiation ☐ 9	Contact with electricity or an electrical discharge ☐ 13
Struck by moving vehicle ☐ 3	Fall from a height* ☐ 7	Exposure to or contact with a harmful substance ☐ 10	Injured by an animal ☐ 14
Struck against something fixed or stationary ☐ 4	*Distance through which person fell ☐ (metres)	Exposure to fire ☐ 11	Other kind of accident (give details in Section H) ☐ 15

Spaces below are for office use only. ☐

F Agent(s) involved – *see note 7*

Indicate which, if any, of the categories of agent or factor below were involved (*tick one or more of the boxes*) –

Machinery/equipment for lifting and conveying ☐ 1	Process plant, pipework or bulk storage ☐ 5	Live animal ☐ 9	Ladder or scaffolding ☐ 13
Portable power or hand tools ☐ 2	Any material, substance or product being handled, used or stored. ☐ 6	Moveable container or package of any kind ☐ 10	Construction formwork, shuttering and falsework ☐ 14
Any vehicle or associated equipment/ machinery ☐ 3	Gas, vapour, dust, fume or oxygen deficient atmosphere ☐ 7	Floor, ground, stairs or any working surface ☐ 11	Electricity supply cable, wiring, apparatus or equipment ☐ 15
Other machinery ☐ 4	Pathogen or infected material ☐ 8	Building, engineering structure or excavation/underground working ☐ 12	Entertainment or sporting facilities or equipment ☐ 16
			Any other agent ☐ 17

Describe briefly the agents or factors you have indicated –

G Dangerous occurrence or dangerous gas fitting – *see notes 8 and 9*

Reference number of dangerous occurrence ☐ Reference number of dangerous gas fitting ☐

H Account of accident, dangerous occurrence or flammable gas incident - *see note 10*

Describe what happened and how. In the case of an accident state what the injured person was doing at the time –

Signature of person making report ☐ Date ☐

Box 2

Tick this box if someone has suffered one of the following major injuries or conditions as a result of an accident arising out of or in connection with work:

- fracture of the skull, spine or pelvis
- fracture of any bone: in the arm or wrist, but not a bone in the hand; or in the leg or ankle, but not a bone in the foot
- amputation of: a hand or foot; or a finger, thumb or toe, or any part thereof if the joint or bone is completely severed
- the loss of sight of an eye, a penetrating injury to an eye, or a chemical or hot metal burn to an eye
- either injury(including burns) requiring immediate medical treatment, or loss of consciousness, resulting in either case from an electric shock from any electrical circuit or equipment, whether or not due to direct contact
- loss of consciousness resulting from lack of oxygen
- decompression sickness (unless suffered during an operation to which the Diving Operations at Work Regulations 1981 apply) requiring immediate medical treatment
- either acute illness requiring medical treatment, or loss of consciousness, resulting in either case from the absorption of any substance by inhalation, ingestion or through the skin
- acute illness requiring medical treatment where there is a reason to believe that this resulted from exposure to a pathogen or infected material
- any other injury that results in the person injured being admitted immediately into hospital for more than 24 h.

Box 3

Tick this box if an employee of yours, a person to whom you are providing training for employment, or a self-employed person working in premises under your control (or you, if you are a self-employed person in your own premises) received an 'over-3-day' injury at work, i.e. an injury that was not one of those listed above, but resulted in incapacity for work for more than 3 days.

Box 4

Tick this box if there has been one of the dangerous occurrences listed overleaf. If a reportable injury was caused you should also tick Box 1, 2 or 3, whichever is appropriate.

Box 5

Tick this box if you are a supplier of flammable gas through a fixed pipe system or a filler, importer or supplier of LPG in a refillable container and you are reporting a fatal injury or one of those listed above, which arose from an incident involving that gas (but not if the incident was connected with a work activity).

Box 6

Tick this box if you are a supplier of flammable gas through a fixed pipe system and you have found out that a connected gas fitting in a user's premises is or has been faulty and dangerous in one of the ways specified by the Regulations.

3 Person or organisation making the report

Just who must report the events covered by Boxes 1–4 in Section A depends upon the circumstances and upon who, if anyone, is killed or injured. The onus to report might be placed on, for example: the employer of an injured person; a self-employed person; someone in control of premises where work is carried out; or someone who provides training for employment. Detailed guidance on this can be found in HSE booklet HS(R)23.

4 Date, time and place

In addition to entering the date, time and address where the reportable event happened (if different from that of the person making the report), you are asked to state:

- precisely where on the premises or site the event happened, e.g. foundry, storeroom, canteen, hospital laundry, grain store, etc.
- what type of work activity is normally carried out there (or was being carried out at the time) as part of your business or undertaking, if not already clear from your description of the place.

5 The injured person

A few examples may help to show what is needed in the 'employment status' and 'trade, occupation or job title' parts of this section. If the injured person was:

(a) a plumber employed by you, then you would tick Box 10 and write 'plumber' in the space provided for trade, occupation, etc.,

(b) a self-employed plumber, then you would tick Box 11 and write 'plumber',

(c) an employee of yours receiving formal training in plumbing either as a YTS trainee or as an apprentice, then you would tick Box 10 and either Box 12 or 13 as appropriate and write 'plumber',

(d) a JTS trainee in joinery, then you would tick Box 13 and write 'joinery',
(e) a school pupil, college student, patient in hospital or customer in a shop, then you would tick Box 14 and write 'pupil', 'student', 'patient' or 'customer' in the space provided for trade, occupation, etc.

6 Kind of accident

If the accident involved a sequence of two or more of the events listed here, then tick the box for the one that happened first (only one box should be ticked in this section).

If the accident did not involve any of these as the primary event, then tick Box 15 and give as much detail about the accident as you can in Section H of the form, after you have completed Section F.

7 Agent(s) involved

Having classified the kind of accident in Section E of the form, you are asked here to describe just what thing or hazard was actually involved—the principal agent or factor.

You should do this firstly by indicating which of the listed broad categories the agent fits into and, secondly, by describing it more precisely in writing—giving its name, type and/or purpose.

You can tick more than one box if more than one of the listed agents was involved—a written description of each should be given. Tick Box 17 if none of the other boxes cover the accident that you are reporting.

Example

If the reported injury is a burn arising from an accident involving the ignition of a flammable liquid escaping from a fractured pipe in a chemical plant, then you should tick Box 11 in Section E and Boxes 5 and 6 in Section F and your written description in Section F should refer to the pipe and its use and to the flammable substance involved.

8 Dangerous occurrences

The list of reportable dangerous occurrences contained in Part I of Schedule 1 of the Regulations is reproduced below. Enter the appropriate paragraph number in the box provided on the form.

Example

For an incident involving the overturning of a road tanker containing petroleum and the escape of petroleum from the tank, enter the paragraph number 13(1)(b)(i).

9 Dangerous gas fittings

A coding system for use when reporting dangerous gas fittings is available to piped gas suppliers on a separate leaflet.

10 Account of accident, dangerous occurrence or flammable gas incident

What is needed here in every case is a description that gives a clear picture of the event being reported and how it happened.

List of dangerous occurrences

Schedule 1, Part I of the Reporting of Injuries, Diseases and Dangerous Occurrences Regulations 1985.

1 The collapse of, the overturning of, or the failure of any load-bearing part of:

(a) any lift, hoist, crane, derrick or mobile-powered access platform, but not any winch, teagle, pulley block, gin wheel, transporter or runway,
(b) any excavator,
(c) any pile-driving frame or rig having an overall height, when operating, of more than 7 m.

2 The following incidents at a fun-fair (whether or not a travelling fun-fair) while the relevant device is in use or under test:

(a) the collapse of, or the failure of any load-bearing part of, any amusement device provided as part of the fun-fair, which is designed to allow passengers to move or ride on it or inside it,
(b) the failure of any safety arrangement connected with such a device, which is designed to restrain or support passengers.

3 Explosion, collapse or bursting of any closed vessel, including a boiler or boiler tube, in which the internal pressure was above or below atmospheric pressure, which might have been liable to cause the death of, or any of the injuries or conditions covered by Regulation 3(2) to, any person, or which resulted in the stoppage of the plant involved for more than 24 h.

4 Electrical short circuit or overload attended by fire or explosion, that resulted in the stoppage of the plant involved for more than 24 h and which, taking into account the circumstances of the occurrence, might have been liable to cause the death of, or any of the injuries or conditions covered by Regulation 3(2) to, any person.

5 An explosion or fire occurring in any plant or place that resulted in the stoppage of that plant or suspension of normal work in that place for more than 24 h, where such explosion or fire was due to the ignition of process materials, their by-products (including waste) or finished products.

6 The sudden, uncontrolled release of one tonne or more of highly flammable liquid, within the meaning of Regulation 2(2) of the Highly Flammable Liquids and Liquefied Petroleum Gases Regulations 1972, flammable gas or flammable liquid above its boiling point from any system, plant or pipeline.

7 A collapse or part-collapse of any scaffold that is more than 5 m high, which results in a substantial part of the scaffold falling or overturning; and where the scaffold is slung or suspended, a collapse or part-collapse of the suspension arrangements (including any outrigger), which causes a working platform or cradle to fall more than 5 m.

8 Any unintended collapse or partial collapse of:

(a) any building or structure under construction, reconstruction, alteration or demolition, or of any false-work, involving a fall of more than 5 t of material,
(b) any floor or wall of any building being used as a place of work, not being a building under construction, reconstruction, alteration or demolition.

9 The uncontrolled or accidental release or the escape of any substance or pathogen from any apparatus, equipment, pipework, pipeline, process plant, storage vessel, tank, in-works conveyance tanker, land-fill site, or exploratory land drilling site which, having regard to the nature of the substance or pathogen and the extent and location of the release or escape, might have been liable to cause the death of, or any of the injuries or conditions covered by Regulation 3(2) to, or other damage to the health of, any person.

10 Any ignition or explosion of explosives, where the ignition or explosion was not intentional.

11 Failure of any freight container or failure of any load-bearing part thereof while it is being raised, lowered or suspended; in this paragraph 'freight container' means a container within the meaning of Regulation 2(1) of the Freight Containers (Safety Convention) Regulations 1984.

12 Either of the following incidents in relation to a pipeline as defined by Section 65 of the Pipelines Act 1962:

(a) the bursting, explosion or collapse of a pipeline or any part thereof,
(b) the unintentional ignition of anything in a pipeline, or of anything that, immediately before it was ignited, was in a pipeline.

13 (1) Any incident:

(a) in which a road tanker or tank container used for conveying a dangerous substance by road:
    (i) overturns
    (ii) suffers serious damage to the tank in which the dangerous substance is being conveyed
(b) in which there is, in relation to such a road tanker or tank container:
    (i) an uncontrolled release or escape of the dangerous substance being conveyed
    (ii) a fire that involves the dangerous substance being conveyed.

(2) In this paragraph, 'conveyance by road', 'road tanker', 'tank container' and 'dangerous substance' has, in each case, the meaning assigned to it by Regulation 2(1) of the Dangerous Substances (Conveyance by Road in Road Tankers and Tank Containers) Regulations 1981.

14 (1) Any incident involving a vehicle conveying a dangerous substance by road, other than a vehicle to which Paragraph 13 applies, where there is:

(a) an uncontrolled release or escape of the dangerous substance being conveyed from any package or container
(b) a fire that involves the dangerous substance being conveyed.

(2) In this paragraph 'dangerous substance' means a substance that is dangerous for conveyance as defined in Regulation 2(1) of the Classification, Packaging and Labelling of Dangerous Substances Regulations 1984.

15 Any incident where breathing apparatus, while being used to enable the wearer to breathe independently of the surrounding environment, malfunctions in such a way as to be likely either to deprive the wearer of oxygen or, in the case of use in a contaminated atmosphere, to expose the wearer to the contaminant,

to the extent in either case of posing a danger to his health, except that this paragraph shall not apply to such apparatus while it is being:

(a) used in a mine
(b) maintained or tested

16 Any incident in which plant or equipment either comes into contact with an uninsulated overhead electric line in which the voltage exceeds 200 V, or causes an electrical discharge from such an electric line by coming into close proximity to it, unless in either case the incident was intentional.

17 Any case of an accidental collision between a locomotive or a train and any other vehicle at a factory or at dock premises that might have been liable to cause the death of, or any of the injuries or conditions covered by Regulation 3(2) to, any person.

chapter three

Industrial studies

After reading this chapter you should be able to:

1. identify the electrical contractor's role within the main structure of the industry,
2. state the purpose of contract documents, specifications, variations, site organisation, good industrial and customer relations,
3. recognise the need for completing various documents, such as daywork sheets and time sheets as well as reports in site diaries,
4. describe various types of drawing and distinguish between various types of diagram, such as block, circuit and schematic,
5. perform measurements from scale drawings and recognise various BS 3939 graphical location symbols on drawings,
6. construct 'as-fitted' drawings from work completed on site.

When a young trainee joins an electrical contracting firm, it is difficult for him at first to appreciate his position and responsibility within that company, particularly if the company is fairly large and employs a large workforce of skilled and non-skilled operatives. It will take him some time to acquire a knowledge of how the company operates but he will eventually come to appreciate the more important role the company plays in the wider industry. Figure 3.1 illustrates where the *electrical contractor* is likely to be found alongside other nominated and non-nominated subcontractors operating under contract conditions to a *main building contractor*.

The main builder will be under the control of an *architect* who is employed by the person having the work done. This person is of course the *client*, who is often the owner of the premises.

When the electrical contractor is tendering (submitting a price) for a project, it is usual for the contract to specify a time for the work to be undertaken. On the successful award of a contract, the electrical contractor may be requested to produce a programme of work, and this will be forthcoming from the plans, drawings and specifications already passed on by the *design team* within the company or by specialist electrical consultants. The *installation team* (comprising electrical foreman, skilled electricians and trainees) will have the task of wiring the installation according to the requirements laid down in the electrical specification or sometimes the designer's schedule.

Contract document

There are various forms of contract used in the construction/engineering industry, such as the JCT Standard Form of Building Contract and the HM Government Form GC/Works/1. Such contracts are assurances from both parties that the work and payment will be carried out in an honest and professional manner; they also provide the necessary protection against sharp practice for both parties. Some of the points raised will be the following.

(a) The work is to be carried out in a workmanlike manner using appropriate materials of good quality.
(b) Qualified labour, paid the correct rates, is to be used.
(c) Adequate supervision is to be provided and the contractor will work to a programme, offer a guarantee for his work and will duly be paid on submission of his interim certificates.

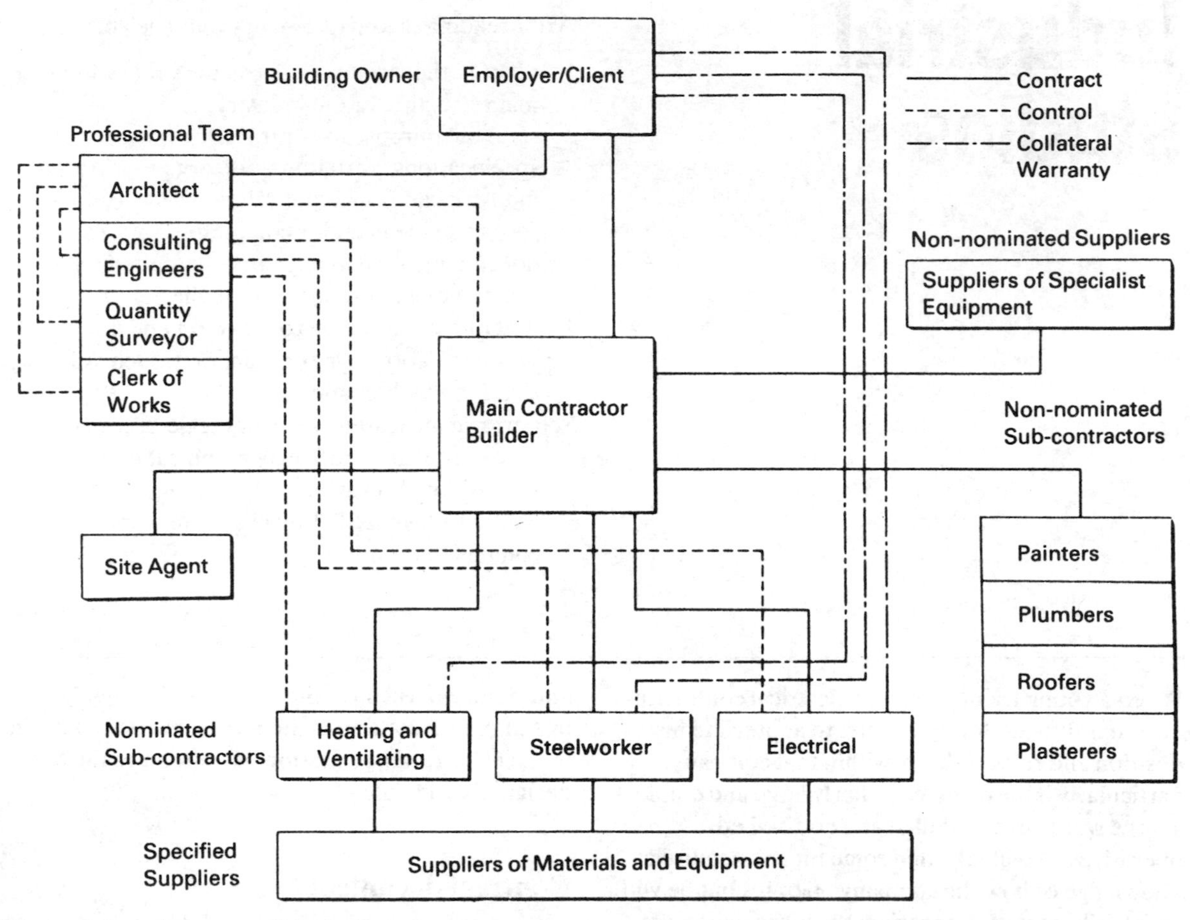

Figure 3.1 *Structure of the industry*

(d) Additional monies will be payable for increased costs should the contract be on a fluctuating basis and not a fixed price basis.

Specifications

Specifications generally consist of two sections, one being the requirements for standards of workmanship as endorsed by Regulation 13–1 of the IEE Wiring Regulations and the other being a specific requirement for the electrical wiring of the installation. Particular requirements are detailed in individual clauses.

In the standard section of the specification, the requirements for such items as conduit will be detailed, and here the electrical contractor will be required to install conduit of the appropriate size, using the correct tools and equipment. He will also be required to ensure that the number of bends between draw-in boxes comply with the requirements of the IEE Wiring Regulations. A typical specification might read as follows:

Specification for electrical installations in dwellings with a floor area not exceeding 200 square metres

Note: This specification is intended for use where there is a public supply of 240 V single phase having the neutral conductor permanently connected to the general mass of earth.

Scope: 1 This specification covers the supply and erection of materials necessary for the complete

electrical installation starting from the termination of the incoming supply mains (intake point) and finishing at:
Lighting points up to and including lampholders and switches
Socket outlets, complete with plugs
Cooker control position
Immersion heater control at hot water tank.

Wiring system: 2 The installation shall be carried out in heavy gauge, screwed welded conduit using screwed fittings and the wiring will be single-core PVC cable.

Main control: 3 At the termination of the supply mains, in the position marked on the plan, the electrical contractor shall supply and fix a consumer's control unit containing:
one 80 A d.p. main switch incorporating a 30 mA r.c.d.
one 45 A m.c.b. for cooker control unit
two 30 A m.c.b.s for socket outlets
one 15 A m.c.b. for immersion heater
two 5 A m.c.b.s for lighting one spare way.

The electrical consultant has the responsibility of ensuring that the specification correctly identifies and details the work that the electrical contractor is to undertake; there will be clauses requiring him to accept responsibility for the satisfactory design of the installation, as well as clauses requiring him to point out any alleged deficiencies or errors in the design at the tender stage of the project. In addition to this, is the general responsibility upon the electrical contractor and his employees under the Health and Safety Act as previously mentioned, and this is to ensure that the electrical installation is safe and designed to a satisfactory standard.

Variations

It sometimes happens that the client changes the requirements about some aspect of the work being undertaken or that an incorrect detail is discovered on a drawing or in the specification. Such *variations* (as they are called) are generally advised by the consulting engineer to the architect who then issues an AI (architect's instruction) to the main contractor who in turn informs the electrical contractor to carry out the extra work in accordance with the terms and conditions of the contract.

Bills of quantities

A bill of quantities is produced by a quantity surveyor and therein is detailed the quantities of all the materials necessary to carry out the works. The contractor has to cost the materials and assess the labour requirement and add on his preliminaries percentage and overhead and profit percentage, thus giving a selling price for each quantity. In many ways a bill of quantities is a more precise method of tendering, as it avoids measuring errors or omissions on the part of the contractor and the various tenders simply reflect the contractor's keenness to win the contract by keeping overheads and profits tight. Some bills may require the contractor to validate quantities and these are not so popular with contractors.

Site organisation

Good site organisation usually reflects an improved productivity, better industrial relations and, therefore, improved profit margins.

From the subcontractor's point of view, he is somewhat in the hands of the main contractor. A well-run building contract will usually inspire the other service contractors to operate efficiently, whereas a poorly operated building contract will result in delays, slow progress and an untidy site with poor facilities. All this can be very demoralising to the workforce.

An electrical contractor is required to have a good site foreman to act as the company's representative on site. He is the basis of all organisation and incentive. The electrical foreman will be responsible for maintaining an adequate supply of material on site and, in the early days of the project, it is usual to keep all material safely locked away in a site hut. The site hut needs to be dry and warm and large enough to allow operatives to change their clothes, as well as to be used as a site office for clerical duties. The site foreman will need to study plans and drawings and keep safe other documents such as a site diary that keeps a record of site meetings, important phone numbers, dates of accidents and damages to material, etc. The foreman must ensure that all documents are tidily filed and that paperwork such as 'daywork' is kept up to date and forwarded to head office regularly. It is essential to keep a daily record of the labour on site, the major work being undertaken and details and reasons for any delays brought about by other trades. For a relatively large project, a *bar chart* may be used to provide immediate visual information on the various work operations being carried out on site. These tasks or activities are time-related from the beginning to the end of the project. Figures 3.2 and 3.3 are typical charts showing building and electrical operations.

Industrial relations

An unhappy workforce is inefficient, unmotivated and this often results in a loss of earnings for the operatives and loss of profit for the electrical contractor; it is a recipe for disaster.

Fortunately, the electrical contracting industry has an enviable record over the years, whereby national strikes have been non-existent. This has been due mainly to the setting up of the *Joint Industry Board* (JIB). The Board is responsible for determining not only wage rates and working conditions but also

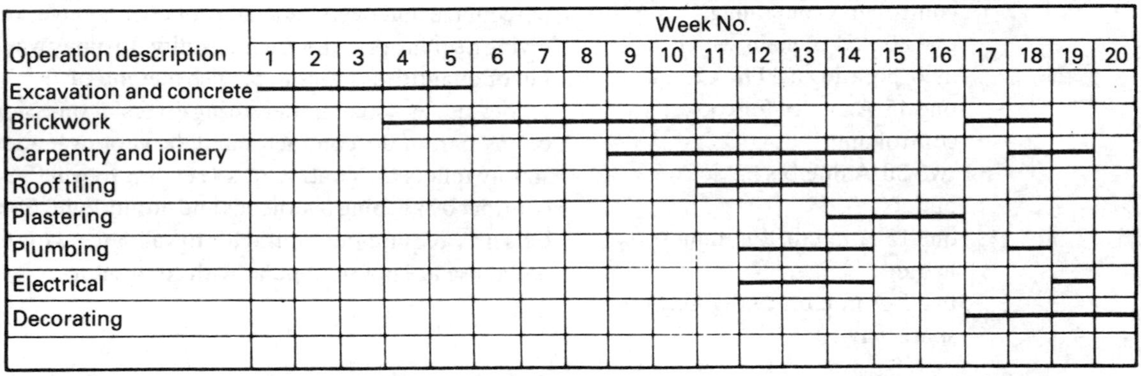

	Week No.																			
Operation description	1	2	3	4	5	6	7	8	9	10	11	12	13	14	15	16	17	18	19	20
Excavation and concrete																				
Brickwork																				
Carpentry and joinery																				
Roof tiling																				
Plastering																				
Plumbing																				
Electrical																				
Decorating																				

Figure 3.2 *Bar chart of simple building trade operation*

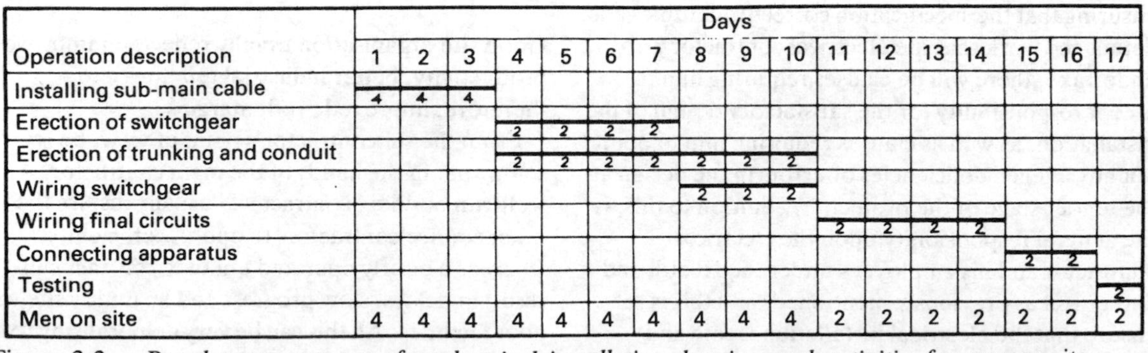

	Days																
Operation description	1	2	3	4	5	6	7	8	9	10	11	12	13	14	15	16	17
Installing sub-main cable	4	4	4														
Erection of switchgear				2	2	2	2										
Erection of trunking and conduit				2	2	2	2	2	2	2							
Wiring switchgear								2	2	2							
Wiring final circuits											2	2	2	2			
Connecting apparatus															2	2	
Testing																	2
Men on site	4	4	4	4	4	4	4	4	4	4	2	2	2	2	2	2	2

Figure 3.3 *Bar chart programme of an electrical installation showing work activities for men on site*

welfare conditions, sickness payment schemes and medical facilities. It has also achieved enormous success with the introduction of its 1983 JIB Training Scheme for junior and senior apprentices. Reference should be made to the JIB working rules and, as a nationally recognised negotiating body, its determinations, once they are made public, are legally binding for employers and employees alike.

The electrical contracting industry is also influenced by its trade union, called the EETPU (Electrical, Electronic, Telecommunication and Plumbing Trade Union). The function of the union is to ensure that its members have a reasonable standard of wage and working conditions, as well as seeing that its members are not victimised by an employer. The Government has set up a body called ACAS (Advisory, Conciliation and Arbitration Service) which can be requested to arbitrate impartially in situations which become deadlocked between employer and employees. An operative who feels that he has been unjustly treated or dismissed by his employer can have his case taken to an industrial tribunal. The tribunal will hear both sides before coming to an impartial decision based on the evidence alone.

Customer relations

From the operatives' point of view, it is unlikely that lengthy meetings with the customer or client would be encountered, but in the event, the operative should be suitably dressed (e.g. in overalls), conscientious and polite. On no account should he make any derisory comments regarding his employer to the client.

The reason for good customer relations is fairly obvious, since a satisfied customer is generally one who will provide future enquiries and possibly be prepared to negotiate the cost of an installation with a known contractor rather than go to a competitive tender.

Many contracting organisations employ staff specially skilled in the art of soliciting enquiries from potential customers. When dealing with overseas clients, there may still be opposition from other countries, and a good knowledge of the language, the various electrical standards, and experience in 'entertaining' and 'putting at ease' will be of invaluable assistance. Many clients are non-technical and rely on agents such as architects or consulting engineers to handle negotiations on their behalf. A presentation of drawings and documents, together with a prompt and efficient service, will be a major advantage to the contractor. In most cases, first impressions count, and a good front-man backed up with an efficient service will win the day.

Records

Electrical contractors have documentation on numerous subjects and the larger companies employ contract engineers and other accounting staff to operate, organise and file the work. The recording of any routine facts and figures will undoubtedly be with the help of *computers* and one has to remember that errors *in* are errors *out*. A computer is only as good as its program and the accuracy of the data input.

In general, correspondence received from different sources must be directed to the member of staff dealing with the subject so that appropriate action may be taken. All correspondence must be suitably filed in a logical sequence, preferably contract/company/date, so as to be readily accessible.

All electrical contractors of any substance will have a number of documents that have to be completed by site operatives. Some of these are as follows.

Daywork sheets

Daywork is the unavoidable work outside the scope of the contract, and has already been dealt with under the section on variations. Briefly, it applies to alterations or extras and will be initiated on the instruction of the architect, engineer or main contractor and, on completion of the work, the operatives/foreman will have to make out a daywork sheet and obtain a signature of approval by the appropriate client's representative. The completed and signed sheet is then returned to the office for pricing up and submission to the client/main contractor for payment (see Figure 3.4).

ALPHA-BETA ELECTRIC COMPANY (U.K.)
Electrical Engineers & Contractors

DAYWORK SHEET

Client ____________________

Job No. ____________________

Date	No. of Men	Start Time	Finish Time	Total Hours	Allow	Notes

Materials

Quantity	Description	Office Use

Supervisor's Signature ____________________

Client's Signature ____________________ Date ____________________

Figure 3.4 *Daywork sheet*

Time sheets

Time sheets are required by office personnel so that operatives' wages can be made up. They provide a continuing record of work completed and time taken or time engaged on site, including any additional payments for travelling, overtime, etc. They are a permanent record of labour on site, should the quantity of labour or rates, etc. be queried at a later date.

Where a bonus system is in progress, operatives will be required to detail time spent on specific tasks and the office staff can then determine any savings on target times and approve any additional payments under the bonus scheme. Figure 3.5 gives typical time sheet details.

Deliveries of material to site

When materials are delivered to site it is imperative that the quantities are checked with the delivery note before receipt of the materials is accepted. To try to obtain additional materials because of short delivery once the supplier is in receipt of a signed delivery note is often impossible. It is therefore most important to see that all materials delivered to site are checked against copies of the original order. If a delivery is

ALPHA-BETA ELECTRIC COMPANY (U.K.)

Electrical Engineers & Contractors

TIME SHEET

Name ______________________________

Week Ending ______________________________

Day	Job No.	Start Time	Finish Time	Total Hours	Travelling Time	Fares Milage
Sun						
Mon						
Tue						
Wed						
Thu						
Fri						
Sat						
Totals						

Employee's Signature: ______________________________

Supervisor's Signature: ______________________________ Date: ______________

Figure 3.5 *Time sheet*

short, then the checking will enable the outstanding material to be chased. Materials broken must also be reported and sent back to the suppliers. It is very important to keep delicate items, such as luminaires and final fixtures, in a separate room or store to avoid damage.

Reports

On matters covering site and office operations, it is advisable that the site foreman or the contract engineer make frequent reports on the progress of the contract. Again, good documentation will enable claims to be easily substantiated, progress to be monitored and action to be taken early, should adverse situations arise that may affect profits.

Difficulties arising on site can be discussed at site meetings with the main contractor. It is at these meetings that variations of work and the orders issued to cover them can be discussed. Any matters that are not cleared should be reported to the electrical contractor's head office.

Interpretation of drawings

Anyone requiring to read and act on information presented to them on drawings will have to be reasonably versed in standard drawing practice. Although many schemes have an accompanying symbols schedule, a reasonable knowledge of BS graphical symbols, as outlined in Chapter 1, is a necessity.

Generally speaking, material drawings are presented in two forms, isometric first-angle and isometric third-angle projection. The angle of view will vary in the way in which the sections and elevations are presented. In some cases, it may be advantageous to give a three-dimensional effect by projecting the layout either isometrically (at 30° to the horizontal) or obliquely (at 45° to the horizontal).

Attention should always be paid to the scale of the drawing, as small details may be of a larger scale for clarity and errors in measuring can easily be made. Some common types of drawing and diagram are given below.

Layout drawings

The architect or the consulting engineer will provide the electrical contractor with building outlines for the purpose of laying out his services. It will usually be the contractor's responsibility to lay out his services and suitably coordinate these with the building layout and other services.

The layout drawings should indicate, with the aid of enlarged details, the actual installation as far as is practicable. These layouts will be required to be modified should any other architect's instruction alter the installation and, in each case, a revision will need to contain a note of the modification together with a revised drawing reference. Figure 3.6 shows a typical layout drawing for part of the wiring system in a building. It will be observed that the scale is 1:50, which means that the drawing is one-fiftieth of the actual size that it represents in real terms. Placing your metric ruler over the length of the drawing and naming 1 cm = 50 cm (or 2 cm = 1 m), the length will be found to be 10.25 m. It is important to remember that this is a plan and allowance has to be made for vertical drops to switches and other controls.

As-fitted drawings

These are the layout drawings modified to the details as marked on the site copies that are maintained as the installation progresses (see Figure 3.7). The revisions will be deleted and the drawings marked '*as fitted*' and copies passed to relevant parties as an accurate record of the installation.

Block diagrams

These are drawings showing the connections between major items of equipment when the internal wiring details would unnecessarily complicate the drawing. Many main distribution schematics are simple line diagrams, as a full schematic diagram would be too difficult to follow and contain unnecessary information. A typical block diagram is shown in Figure 4.16.

Circuit and wiring diagrams

These are the types of diagram generally found, which indicate the actual conductors used and their supply polarities. There are many examples of these circuits within the text, such as that shown in Figure 4.9. For an example of a schematic diagram, see Figure 5.35. In technical college studies, it is very important for students to use the correct drawing aids and to fully label their diagrams. Students should not mix circuit diagram symbols with graphical location symbols.

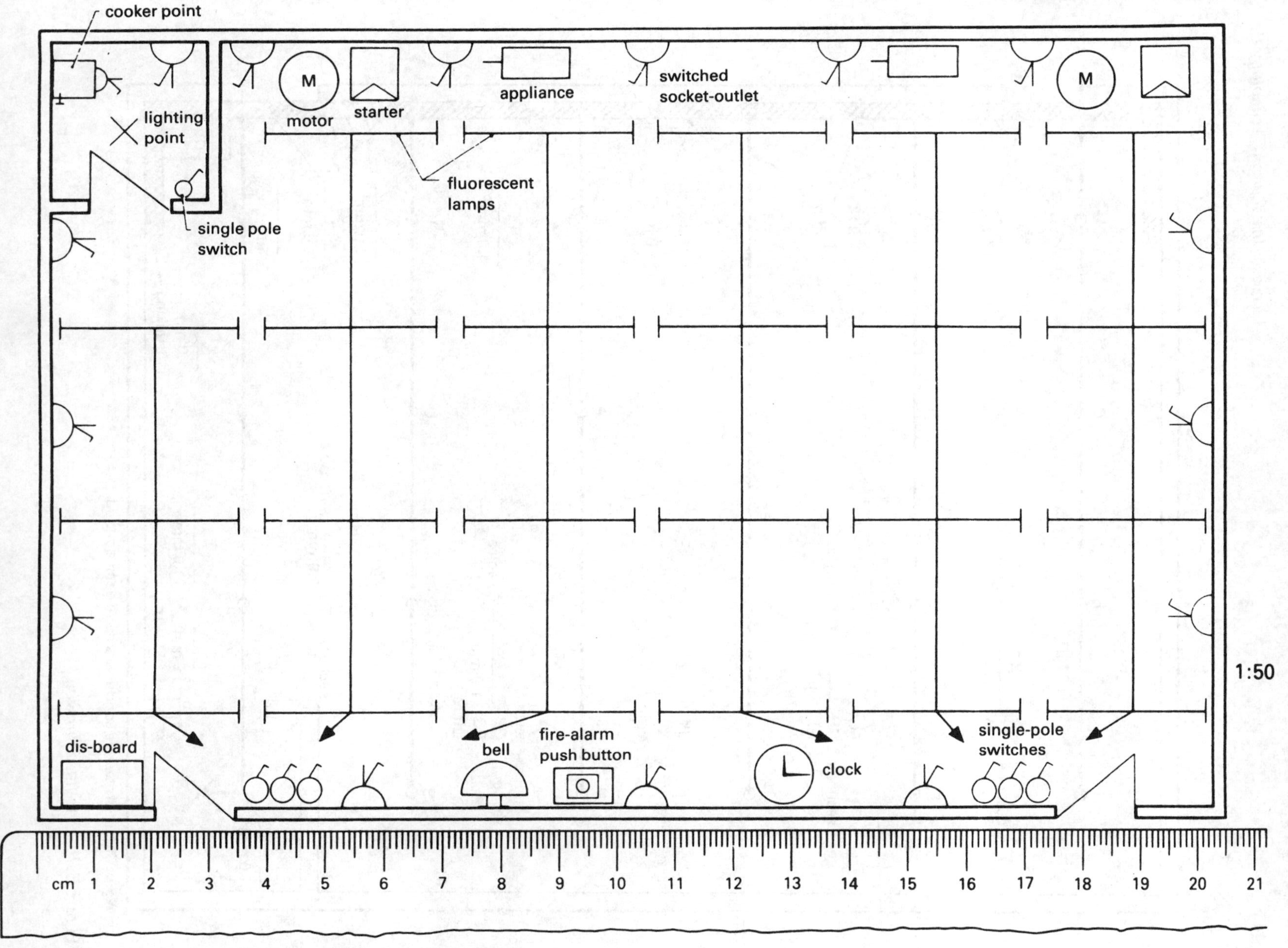

Figure 3.6 *Layout drawing showing electrical requirements*

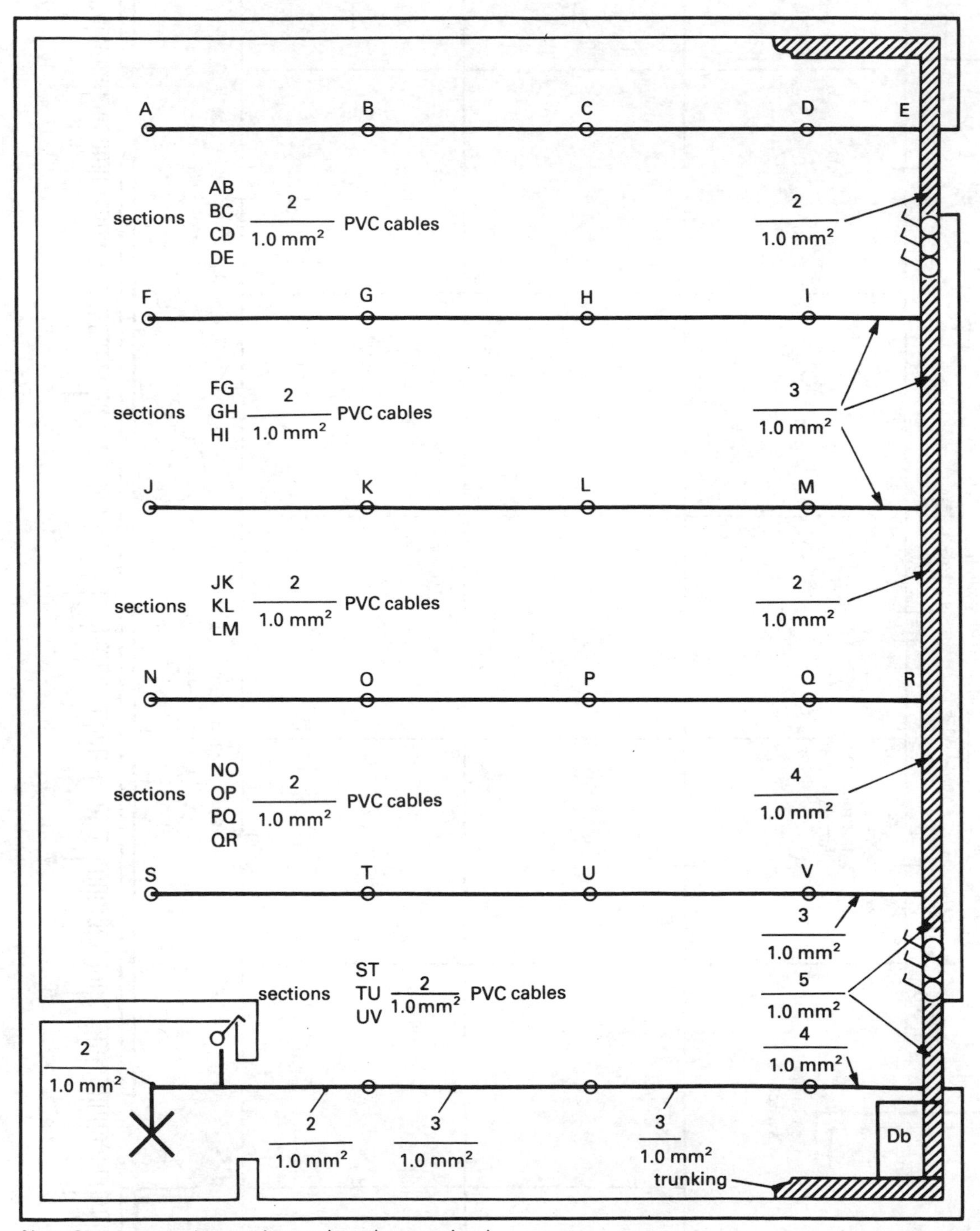

Figure 3.7 *'As-fitted' drawing*

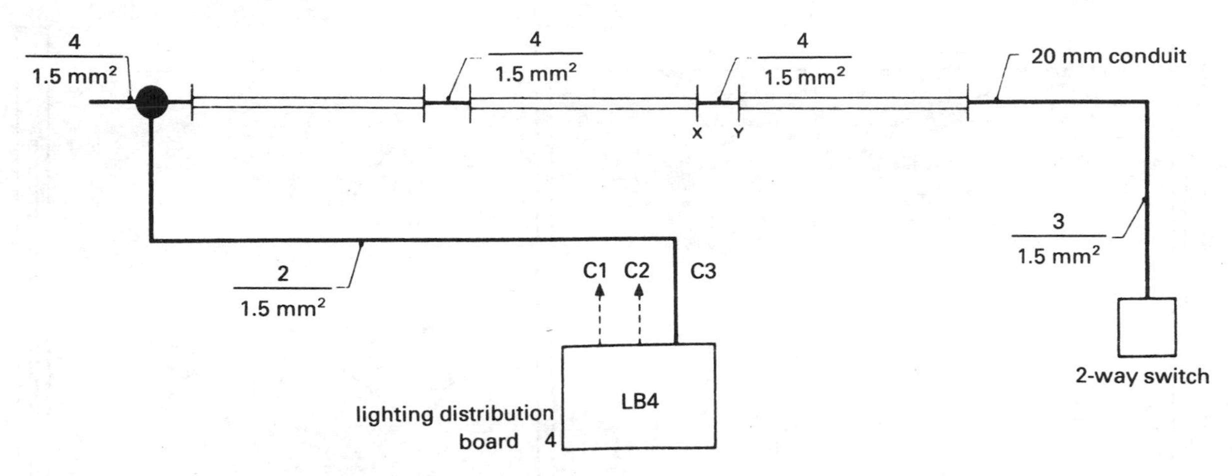

Figure 3.8 *Conduit route and cable size for lighting*

General

Modern drawing practice is to produce drawings in ink on paper negatives, which can be reproduced easily via a photographic print process. Ink drawings require a high practical skill, although general layouts are fairly straightforward to draw. Figure 3.8 shows the arrangement for identifying a typical lighting circuit on a drawing with its associated distribution board. In section XY, four 1.5 mm cables will pass through the conduit (excluding the c.p.c.). The 'as-fitted' drawing mentioned previously is similarly marked.

Exercise 3

1 Figure 3.9 shows a diagram of a two-bedroom flat. The floors and ceilings are of concrete construction and the walls are brickwork with a plaster finish. If the wiring system is single-core PVC cables in 20 mm metric conduit, (a) determine the approximate amount of conduit required, and (b) indicate the number of cables in each section of the conduit, assuming the conduit is not satisfactory as a protective conductor.

Note: The scale is 1:30.

CGLI Part I Assignment 1986 (Modified)

2 Figure 3.10 shows the arrangement for controlling a lighting point using BS 3939 graphical location symbols. Draw the circuit diagram as would be wired from a protective device inside a distribution fuseboard and also show the connections of the circuit protective conductor at each outlet point.

3 Draw a bar chart of a simple project similar to that in Figure 3.3 but for one pair of operatives. Question 2 is a good example to attempt.

4 What advice should operatives be given when working on a mobile tower scaffolding, as shown in Figure 3.11?

5 Prepare notes on the following topics: (a) clocking on and off routines; (b) tea breaks; (c) washing facilities; and (d) site meetings.

6 Briefly explain the procedure on site for: (a) delivery of material; (b) storage of material that is easily damaged; and (c) information required in a site diary.

7 Draw a *circuit diagram* showing the sequence of control into a domestic consumer's premises, starting with supply cut-out fuse and ending with distribution fuseboard supplying final circuits.

8 (a) Using relevant guidance notes, determine the maximum number of single-core PVC cables

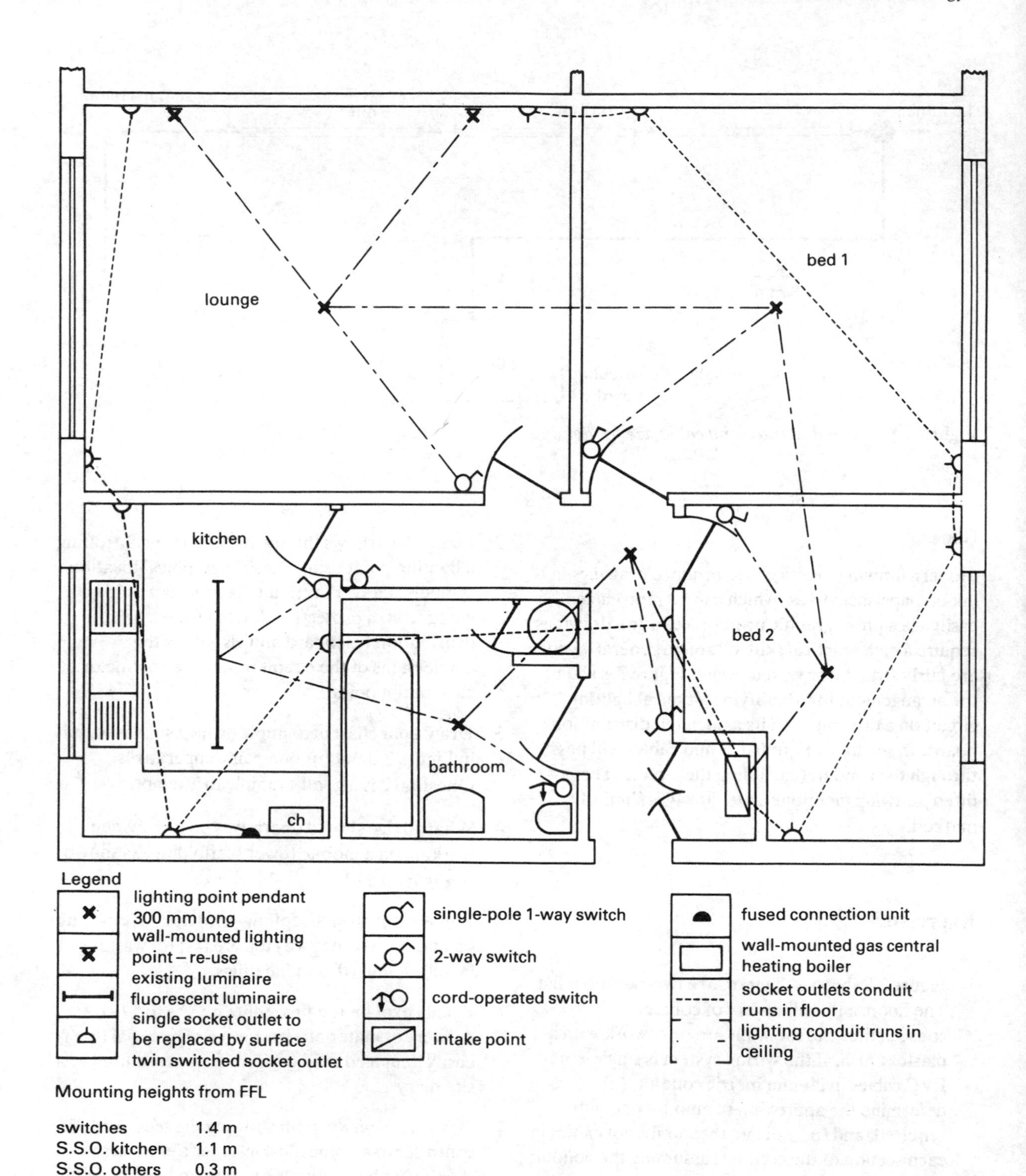
lounge
bed 1
kitchen
bed 2
bathroom
ch
Legend
lighting point pendant 300 mm long
wall-mounted lighting point – re-use
existing luminaire
fluorescent luminaire
single socket outlet to be replaced by surface twin switched socket outlet
single-pole 1-way switch
2-way switch
cord-operated switch
intake point
fused connection unit
wall-mounted gas central heating boiler
socket outlets conduit runs in floor
lighting conduit runs in ceiling
Mounting heights from FFL
switches 1.4 m
S.S.O. kitchen 1.1 m
S.S.O. others 0.3 m
consumer's unit 1.0 m

Figure 3.9 *Diagram of a two-bedroomed flat*

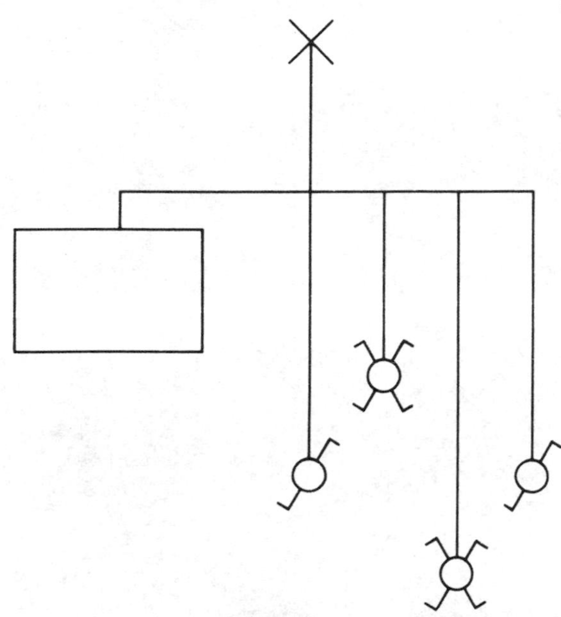

Figure 3.10 *Switching arrangements for a lighting point*

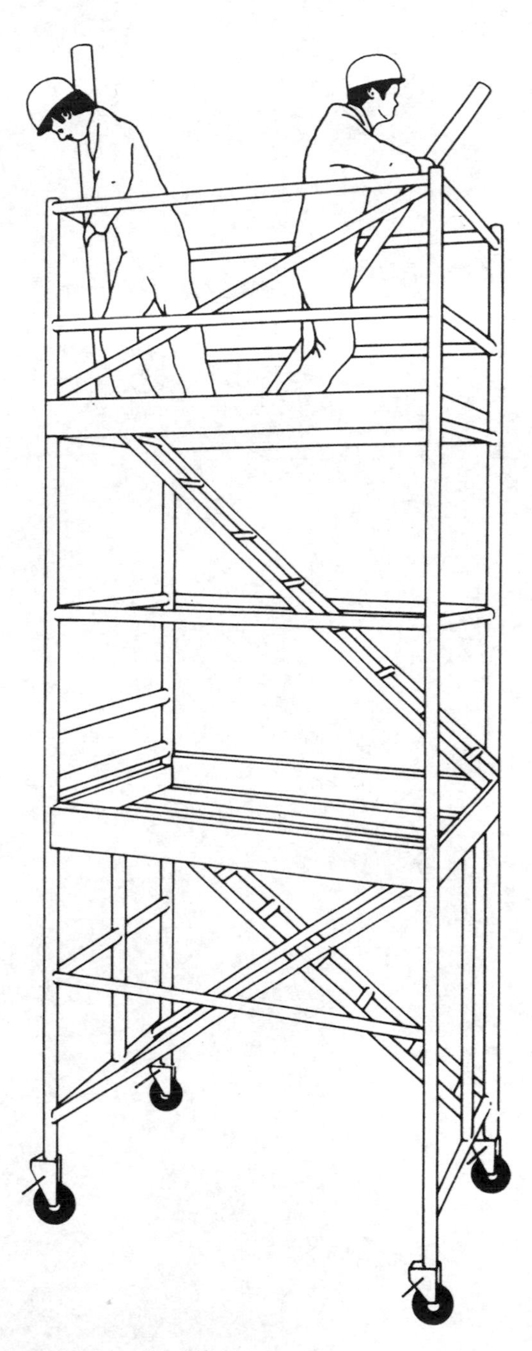

Figure 3.11 *Mobile tower*

allowed in 20 mm metric conduit for the following conditions:

(i) length 4 m; two bends; 1.5 mm^2 cable
(ii) length 9 m; one bend; 2.5 mm^2 cables.

(b) Determine the size of metric conduit to allow six 1.5mm^2 cables and eight 2.5mm^2 cables to be installed. Assuming the worst length of run is 6 m and incorporates two bends.

9 List some of the precautions that you think are necessary when burying armoured cables in an open trench before it is filled in.

10 From relevant manufacturers' literature, select electrical items for the following, quoting the catalogue numbers:

(a) one 13 A switched socket, outlet
(b) five single-pole, one-gang, one-way switches
(c) one 80 A seven-way, metal-clad consumer unit
(d) two 6 A m.c.b.s; one 15 A m.c.b.; three 30 A m.c.b.s; and one 45 A m.c.b.
(e) one 85 W (1800 mm) single fluorescent luminaire.

chapter four

Installation 1

After reading this chapter you should be able to:

1 state the different types of electricity supply given to consumers' premises,
2 state a number of requirements from the IEE Wiring Regulations relating to common earthing systems, such as TN–S, TN–C–S, and TT systems and also the requirements concerning main bonding and supplementary bonding,
3 describe the operation of a residual current device,
4 perform calculations on assessing current demand in designing installations,
5 describe the route taken by a fault current occurring on a consumer's premises,
6 state a number of requirements from the IEE Wiring Regulations with regard to prospective fault current and protection against overcurrent,
7 state a number of common circuit protective devices and understand graphs of their time–current characteristics,
8 perform calculations to find circuit conductor sizes.

Supplies into consumer premises

In England and Wales there are twelve Regional Electricity Companies (RECs) who purchase bulk supplies of electricity and sell it to various types of consumer in the form of retail tariffs. The purpose of these tariffs is to:

(i) recover the cost of energy purchased and also recover the distribution costs involved in making supplies available to consumers,
(ii) encourage an efficient utilisation of electricity through appropriate costing measures,
(iii) provide equity of treatment between one consumer and another.

Each REC offers various types of tariff to its consumers, who may be *domestic*, *commercial* or *industrial*. Most domestic consumers are charged a *two-part tariff* in which there is a single rate of charge for every unit (kWh) of energy used, and a standing charge that is usually payable quarterly (see Figure 4.1). It will be noticed from this customer's bill that there are two unit charges, one at 5.16 p per unit and the other at 1.9 p per unit. This latter charge is based on an *Economy 7 Tariff* and both readings are taken from a dual-rate energy meter and time switch provided by the local REC. These kWh-meters with their '*off-peak*' facility provide 7 h of cheap night-time electricity for the consumer.

Another type of tariff is the *Maximum Demand Tariff* applicable to commercial and industrial users. It consists of a measured maximum demand charge and a unit charge as described above, but the demand charge is based on the customer's monthly maximum demand (in kW or kVA), which is lower in the summer months than in the winter months.

Large industrial users are often provided with their own link from the grid because of their energy consumption and generally complex layout. Some large commercial premises such as a district hospital, a large shopping precinct or even an office block, will

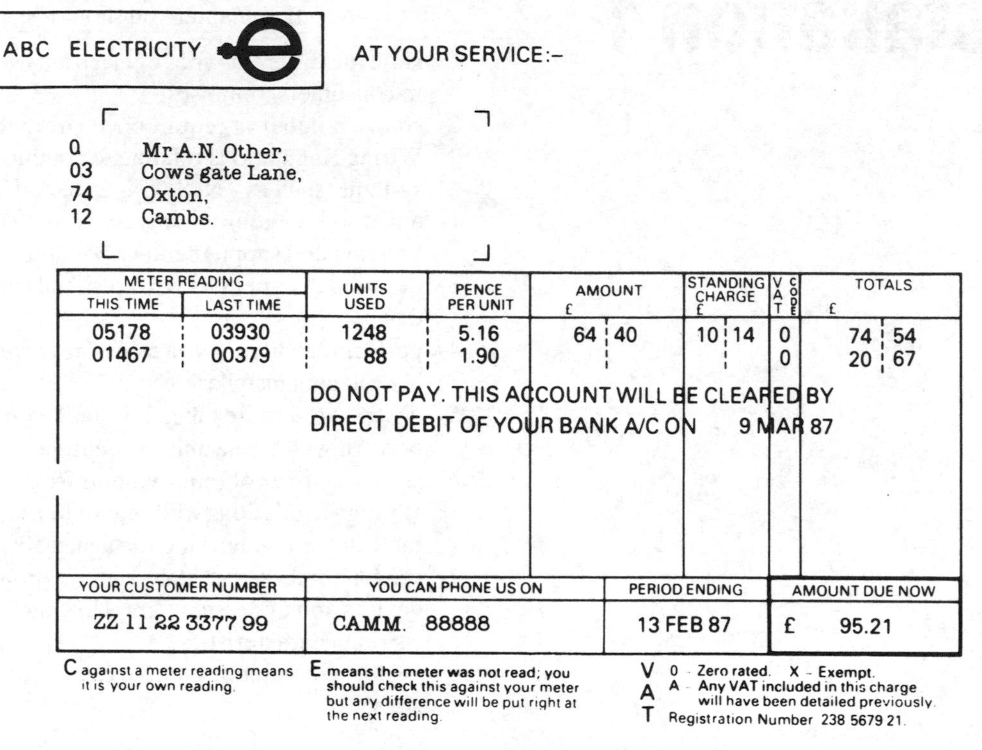

ABC ELECTRICITY AT YOUR SERVICE:–

0 Mr A.N. Other,
03 Cows gate Lane,
74 Oxton,
12 Cambs.

METER READING THIS TIME	METER READING LAST TIME	UNITS USED	PENCE PER UNIT	AMOUNT £	STANDING CHARGE £	VAT	CODE	TOTALS £
05178	03930	1248	5.16	64 40	10 14	0		74 54
01467	00379	88	1.90			0		20 67

DO NOT PAY. THIS ACCOUNT WILL BE CLEARED BY DIRECT DEBIT OF YOUR BANK A/C ON 9 MAR 87

YOUR CUSTOMER NUMBER	YOU CAN PHONE US ON	PERIOD ENDING	AMOUNT DUE NOW
ZZ 11 22 3377 99	CAMM. 88888	13 FEB 87	£ 95.21

C against a meter reading means it is your own reading.

E means the meter was not read; you should check this against your meter but any difference will be put right at the next reading.

VAT 0 – Zero rated. X – Exempt. A – Any VAT included in this charge will have been detailed previously. Registration Number 238 5679 21.

Figure 4.1 *Domestic consumer's quarterly bill*

be provided with a supply of 11 000 V. They may have their own indoor substations where the voltage will be transformed down to 415 V and 240 V by means of a step-down transformer (see Figure 4.2). These lower voltages are also standard for other types of factory or commercial premises, as well as for shops and domestic dwellings. A circuit diagram of the arrangement for reducing the high-voltage a.c. supply down to low voltage is shown in Figure 4.3.

It is worth noting that supply authorities are permitted under the Electricity Supply Regulations 1988 to allow a consumer's voltage to vary by not more than ± 6% and the supply frequency by not more than ± 1%. This means of course that for a domestic consumer supplied at 240 V, 50 Hz, the voltage may vary between 225.6 and 254.4 V, while the frequency is allowed to vary between 49.5 and 50.5 Hz. These limitations are very important, since electrical equipment is specifically designed for operation on standard voltages and at standard frequencies. If the voltage into a consumer's premises is too low, electrical equipment will not function efficiently; if it is too high, lamp filaments and appliances with resistive elements will have shorter life expectancies. Too much variation in supply frequency can also upset a whole range of timing devices that flood the present appliance market.

Where a consumer requires an a.c. supply of 415 V/240 V, it is most likely that the supply authorities will provide him with a *three-phase, four-wire system.* This will allow the electrical contractor, who will probably be the designer and installer, the opportunity to *balance* the final circuit loads over all three phases of the supply so that each phase of the main cable will carry equal current. In theory, this arrangement is the aim of all good design planning, but it may not quite be achieved in practice since single-phase loads are continually being switched on and off at various times. It is more than likely that any one phase of the three-phase supply will be carrying a different current to the other phases. It is for this reason that a neutral conductor is used to carry back

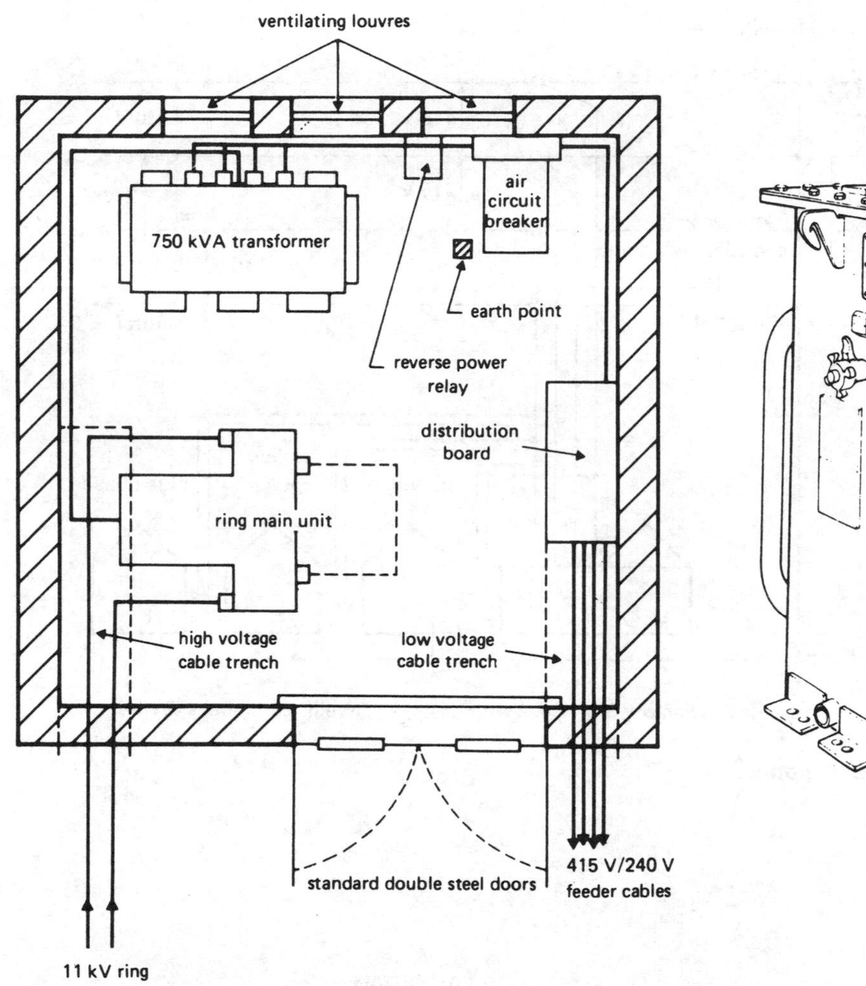

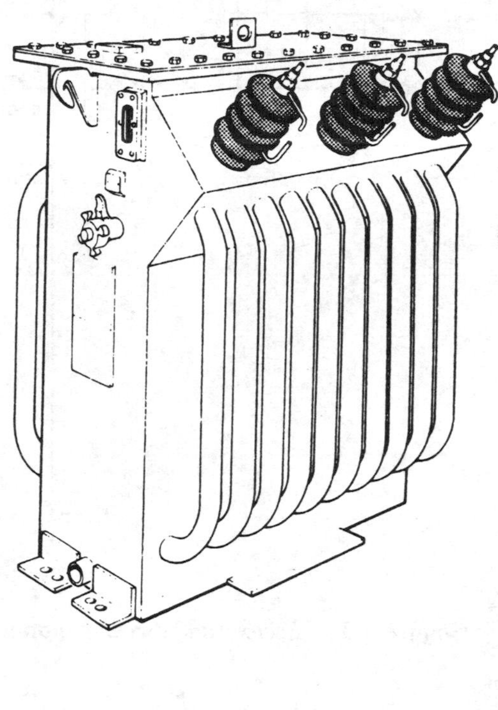

Figure 4.2 *Three-phase 11 kV/415 V transformer*

any *unbalanced* current to the supply transformer. An example of estimating the size of main cable and switchgear is given later in the chapter.

Figure 4.4 shows several methods of supplying consumers' premises. Diagram (a) is a TN–C–S earthing system. The energy meter is the local REC's responsibility and so too is the main cut-out fuse. The electrical contractor will have the responsibility for wiring the installation. This includes earthing the premises as well as earth bonding the gas and water services.

Diagram (b) is another common type of earthing arrangement, called a TN–S system, which shows the added facility of the REC's time switch used to control an *Economy 7 Tariff* charge incorporated inside a dual-rate energy meter. There is a similar meter arrangement that gives both an 'off-peak' supply during the night and a 24-h supply, but this requires separate wiring, as in diagram (c). This latter arrangement is used for heavily loaded circuits such as night storage heaters. Diagram (d) shows the meter connections for a three-phase supply. For simplicity, the earth bonding has been omitted.

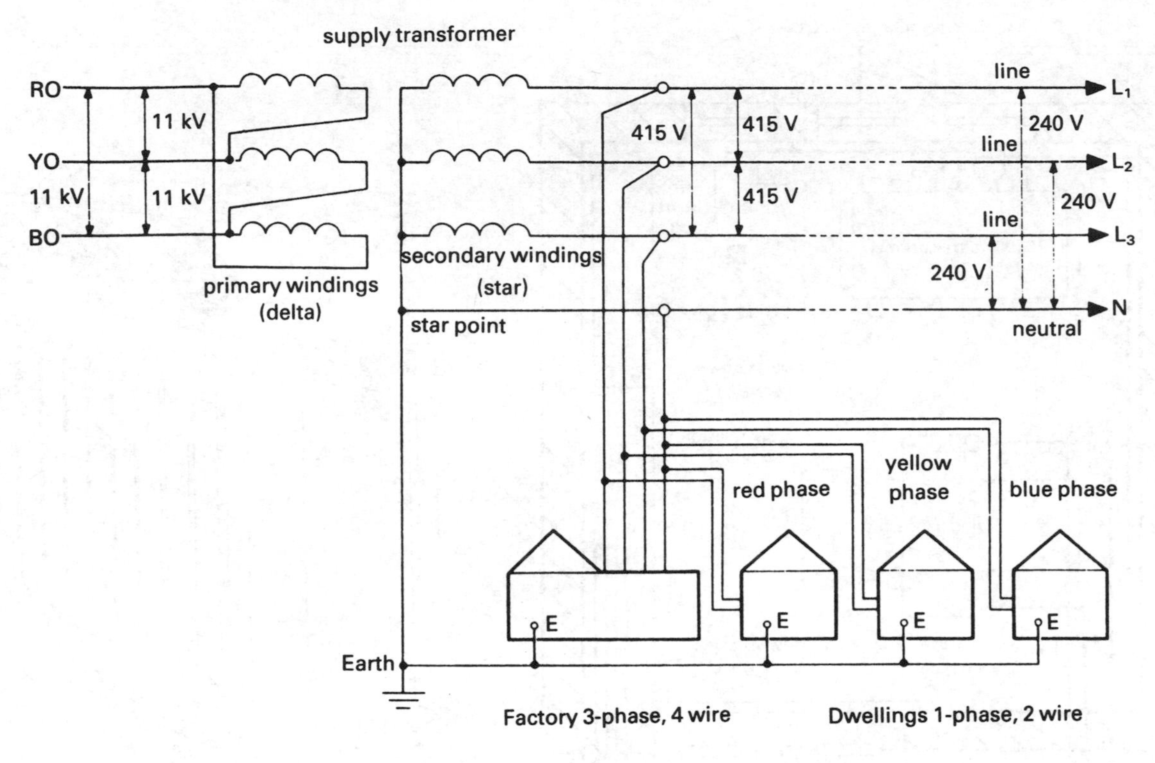

Figure 4.3 *Factory and dwelling supplies*

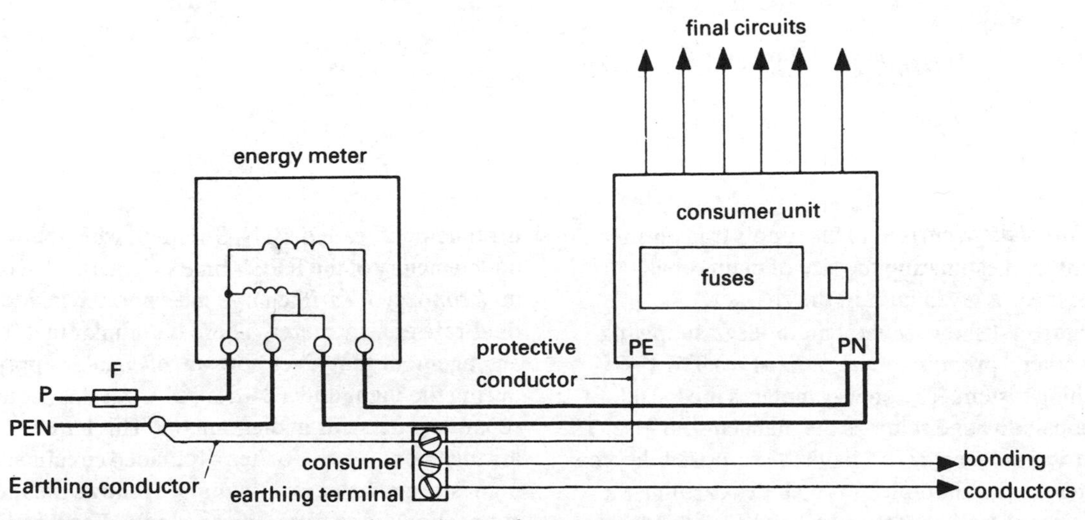

(a) Metering arrangements for TN–C–S supply to a domestic consumer

Figure 4.4 *Supplies to consumers' premises*

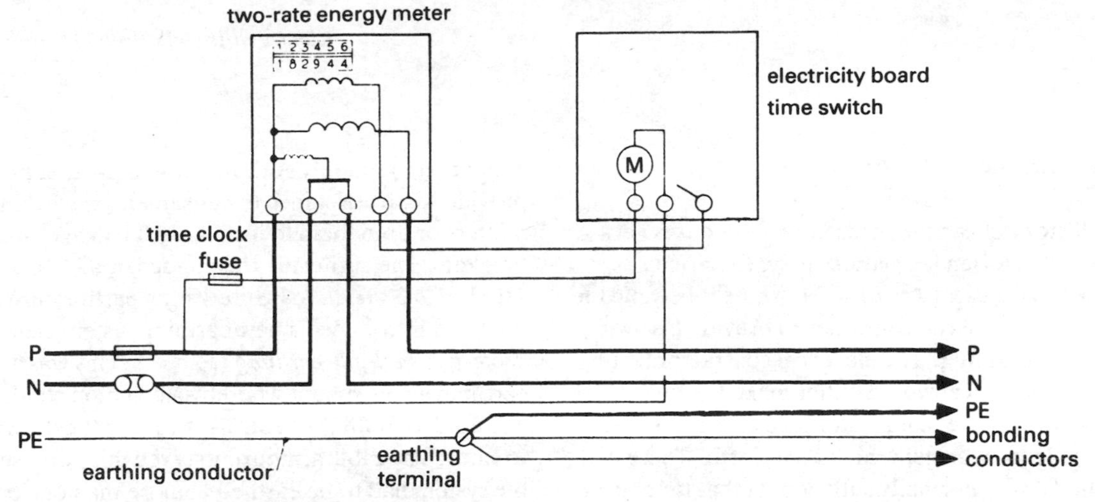

(b) Meter arrangements for Economy 7 tariff

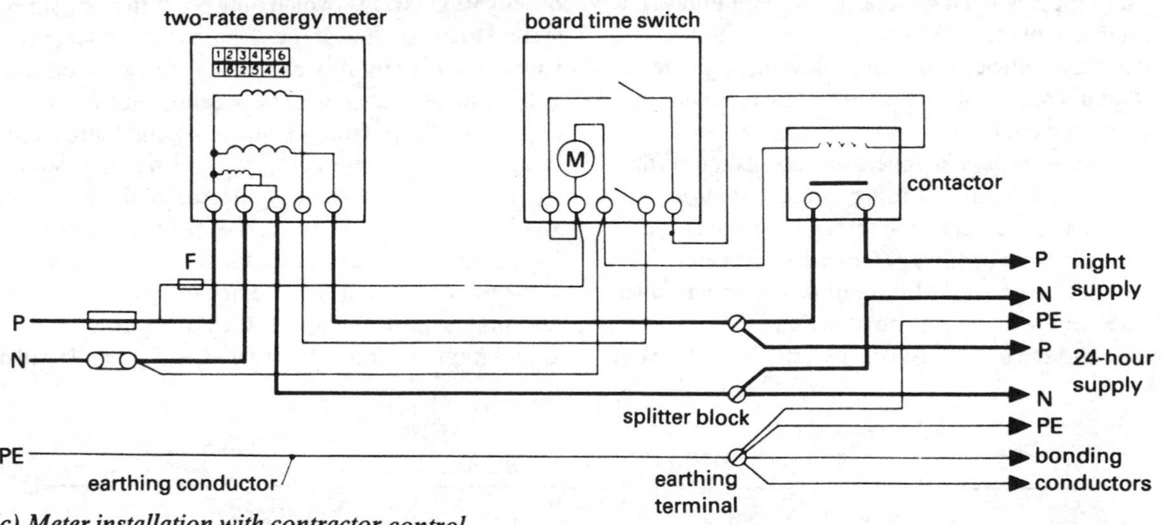

(c) Meter installation with contractor control

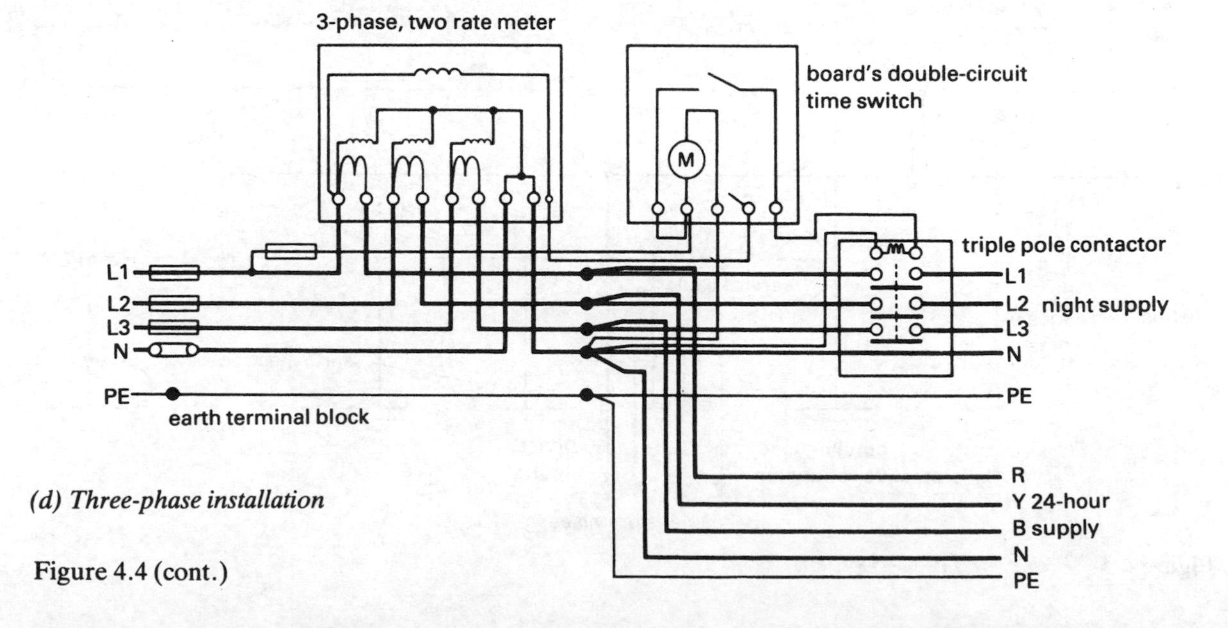

(d) Three-phase installation

Figure 4.4 (cont.)

Earthing arrangements

The history of earthing consumers' premises goes back to Victorian legislation in the Electricity Lighting (Clauses) Act of 1899, where it became the responsibility of the consumer to provide his own means of earthing. The electrical contractor (acting for the consumer) would either make the final earth connection to the main water pipe or drive an earth electrode into the general mass of earth. These two methods were normally sufficient at that time; they created a *third path* for leakage currents to flow back to the supply when a fault occurred. Hopefully, the impedance of this third path was low enough to encourage sufficient current to flow through the soil, so that it would rupture the protective fuse and isolate the faulty circuit.

As the years passed, it became increasingly difficult to provide adequate earthing. More and more consumers were being connected to supplies and difficulties were being experienced in rural areas where the moisture content of the soil was often a problem. Added to this came the introduction of plastic water mains and the use of insulated joints. It finally came about that the supply authorities would either have to provide some assistance to consumers on earthing matters or urge them to install earth leakage circuit breakers. The authorities responded by allowing their supply *cable sheath* to be used as an earth return and they also introduced a new earthing system called *protective multiple earthing* (p.m.e.). This was an earthing arrangement whereby a common conductor was used for both the neutral return and the third path to earth. The cable armouring served this purpose and the system had to be earthed at more than one point (see Figure 4.5). This departed from the Electricity Supply Regulations, which only permitted the supply system to be earthed at the *star point* of the secondary winding of the supply transformer. The method was, at first, somewhat slow to be accepted but several approvals allowed it more flexibility and there were a number of advantages created by using this system. There was one less cable conductor, making significant savings in the costs of the distribution, including savings in termination techniques. Today, it is the most common method of earthing new consumers' premises and in the IEE Wiring Regulations it is known as a *TN–C–S System*. It will be

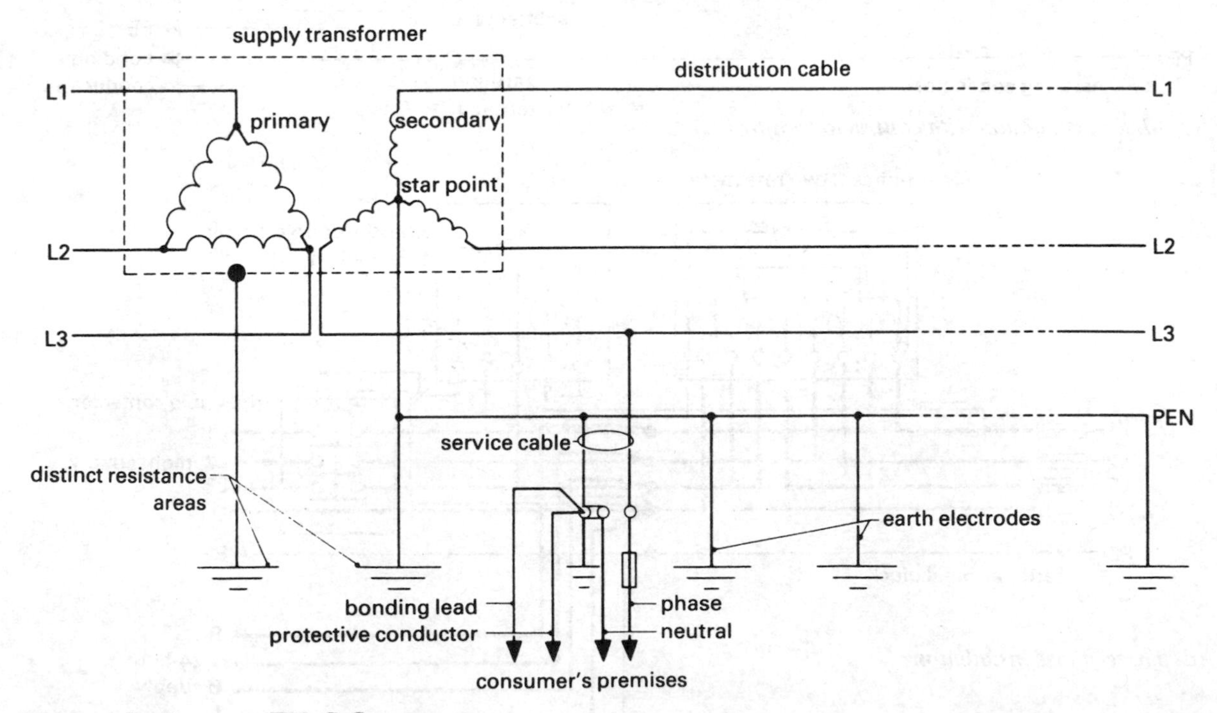

Figure 4.5 *P.m.e. or TN–C–S system*

seen in Figure 4.5 that the common neutral/protective conductor is called a *PEN conductor*. A typical p.m.e. concentric service cable for a single-phase supply is shown in Figure 4.6.

Figures 4.7 and 4.8 show the intake arrangements for TN–S and TT systems.

Figure 4.6 *P.m.e. concentric service cable*

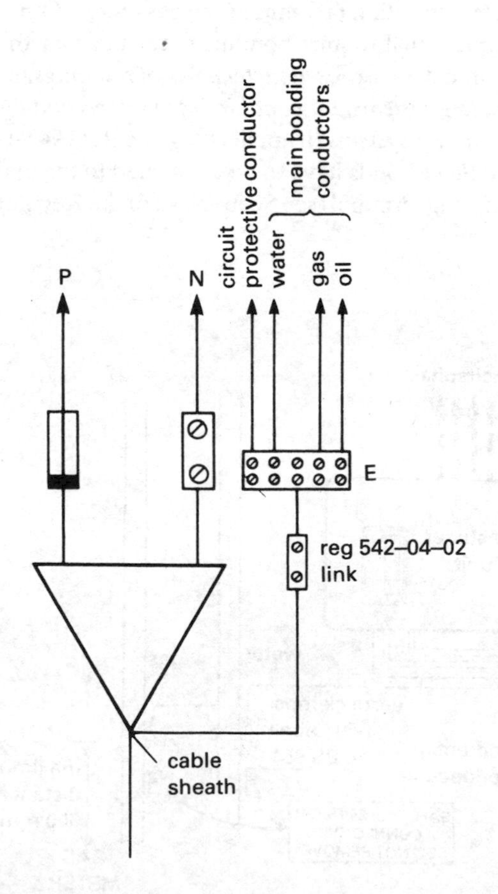

Figure 4.7 *TN–S system*

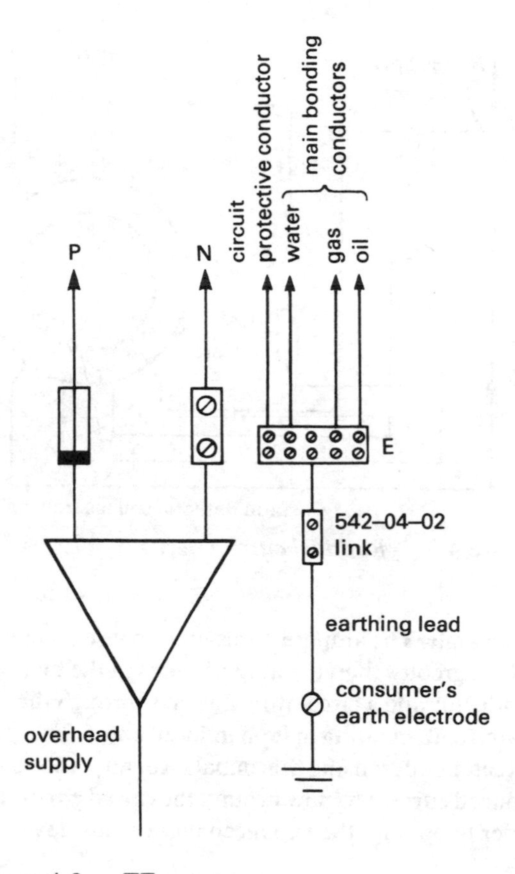

Figure 4.8 *TT system*

The preferred method of detecting earth leakage to avoid the risk of electric shock as well as possible fire risk is called the residual current device. This device is necessary for a TT system. Figure 4.9 shows how the device is connected. It will be seen that the phase and neutral conductors pass through a toroidally wound transformer on their way to the load. Also on the toroid is a search coil connected in series with a trip coil. There is also a test circuit to allow the device to be tested periodically (see IEE Reg. 514–12–02). If the load circuit is healthy and the main switch closed, current will create a magnetic flux in both limbs of the toroid and because the windings on the toroid are equal and opposite, the core flux becomes balanced (i.e. no magnetic flux circulation). The only way for this to happen is by leakage current escaping to earth, as in the case of an

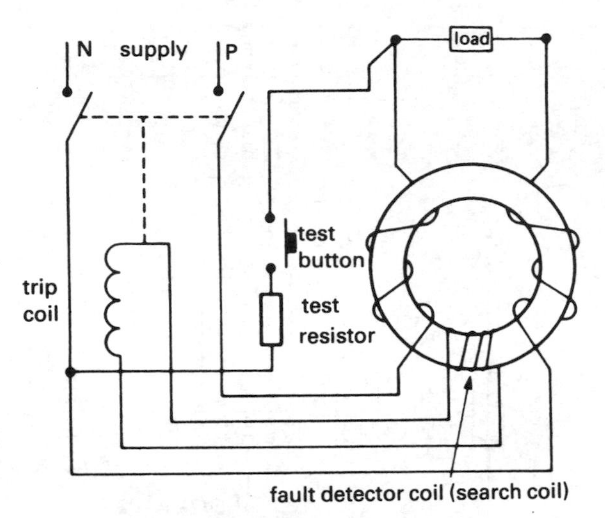

Figure 4.9 *Residual current device*

earth fault. The ampere-turns in the phase conductor will be greater than the ampere-turns in the neutral conductor and a circulating flux cuts through the search coil, creating in it an induced e.m.f. The detector coil, sensing this imbalance, now allows an induced current to flow around the closed circuit in order to operate the trip mechanism of the device.

The r.c.d. needs to satisfy the requirements of Reg. 412–06 and the numerous other regulations contained in Section 531 of the IEE Wiring Regulations.

With regard to the subject of *bonding*, reference should be made to Section 547 of the IEE Wiring Regulations. Here, it is stated in Reg. 547–02–01 that the minimum size of the main equipotential bonding conductor should be not less than 6 mm^2. Where a TN–C–S earthing system is provided and the supply neutral is not greater than 35 mm^2, electricity boards will require 10 mm^2 as the minimum size, but one should always check with the local REC on this matter. The next regulation, Reg. 547–02–02, is also important and concerns the connection of the bonding conductors to the *gas and water services* (see Figure 4.10). These have to be on the consumer's side and as near to the point of entry as possible. It is recommended that the gas service connection should be made within 600 mm of the gas meter. Other services that require bonding to the main earthing terminal are *exposed metal parts of buildings, central heating systems, air conditioning systems and lighting protective systems* (if applicable), see Reg. 542–04–01. All these bonds have to be connected to the main earthing terminal (see Section 542 of the Regulations)

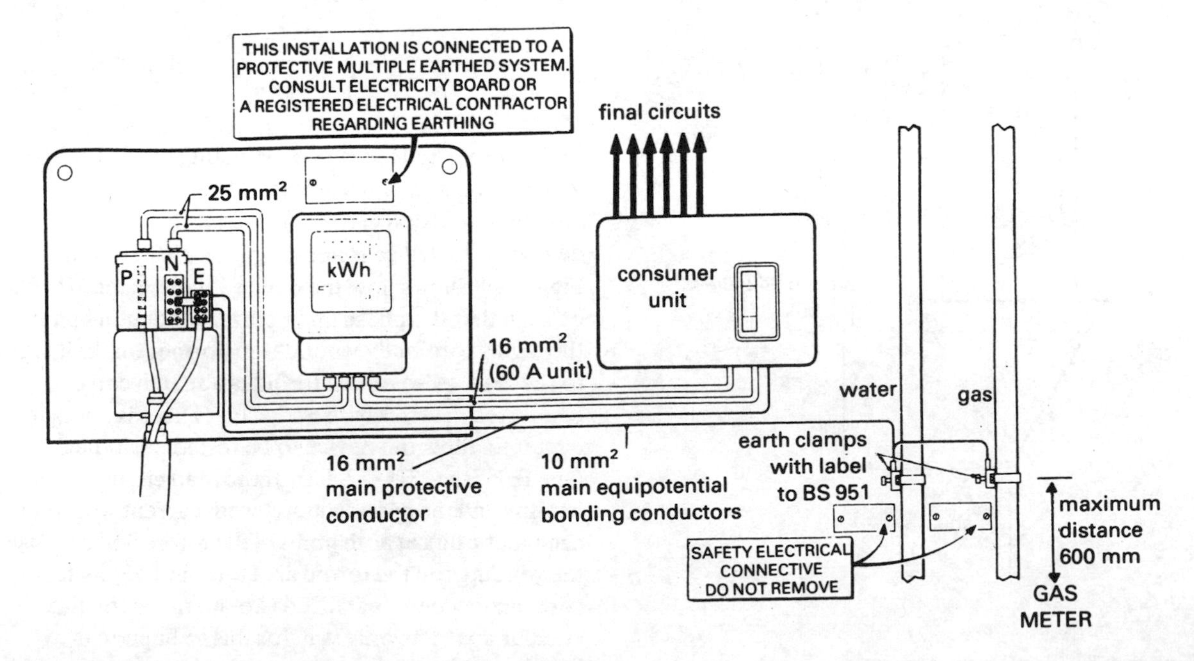

Figure 4.10 *Equipotential bonding of other services*

and provision needs to be made for a test joint in the main earthing conductor in order to carry out resistance measurements tests (see Figure 4.8). This joint must be mechanically and electrically sound and disconnected by means of a tool. At the point where an earthing conductor is connected to an earth electrode and at points where bonding conductors connect to extraneous conductive parts, permanent labels have to be fitted and they should be durably marked with the words: '*Safety Electrical Connection—Do Not Remove*.' This is a requirement of Reg. 514–13–01 and Reg. 542–03–03. A BS 951 earth clamp is often used for this purpose, as shown in Figure 4.11, and is often made of a non-ferrous material such as phosphor bronze or brass.

Figure 4.11 *BS 951 earth clamp*

On the final circuit side of the installation, *supplementary bonding* is required. This is because there are many extraneous conductive parts like pipes, taps, sinks, baths, showers and radiators, etc. connected with plastic washers and the plumbing is not continuously connected back to the service entry point where the main equipotential bonding was carried out. The requirements dealing with these supplementary bonding conductors can be found in Reg. 547–03, which applies to local supplementary bonding to maintain the equipotential zone for those items of metalwork previously mentioned. Metal benches found in a commercial kitchen can transmit an earth potential if they are installed on a tiled floor likely to become wet or if they come in contact with electrical equipment or electrical wiring. Bonding connections to movable metal benches and tables should be made using flexible cables of adequate length to avoid discontinuity; such connections and joints must be made with the use of a tool. It is only acceptable for bonding conductors to be disconnected by electrically skilled persons. Also, Reg. 601–04 requires bonding for all simultaneously accessible exposed conductive parts in a bathroom or shower room (see Figure 4.12). This bonding is required regardless of whether or not any of the parts are connected together by bonding outside of the room and can be done inside an airing cupboard or under the floor. Further requirements are mentioned in Reg. 547–03. In this regulation the stated minimum sizes for supplementary bonding conductors are: 40 mm^2 if there is no mechanical protection given to the cable and 2.5 mm^2 if there is mechanical protection. Where the conductor is exposed, there is more risk of damage, either physical or from corrosion, in a damp environment. Regulation 601–04–01 requires bathroom circuits to be disconnected in 0.4 s but does not apply to SELV circuits.

Whilst it is the prerogative of the supply authorities to tell consumers the minimum size of the main equipotential bonding conductor, the size of the main protective conductor is the choice of the consumer in so far as meeting the requirements. This can be achieved in two ways, either by *calculation* using the formula provided in Reg. 543–01–03 or by using *Table 54G* provided in Reg. 543–01–04. If the former method is used it is possible to select a smaller cable, whereas if the latter method is used it would have to relate to the size of the main phase conductor, e.g. if the meter tails were 10 mm^2 then the protective conductor need only be 10 mm^2; if the meter tails were 25 mm^2 then the protective conductor would have to be 16 mm^2 and if the meter tails were over 35 mm^2 then the protective conductor need only be half the size of the meter tails. This method is easier to apply provided that the protective conductor is made of the same material as the phase conductor.

A consumer's earthing arrangements must also consider *circuit protective conductors* and *earth electrodes*.

Circuit protective conductors are necessary for those final circuits requiring to be earthed and

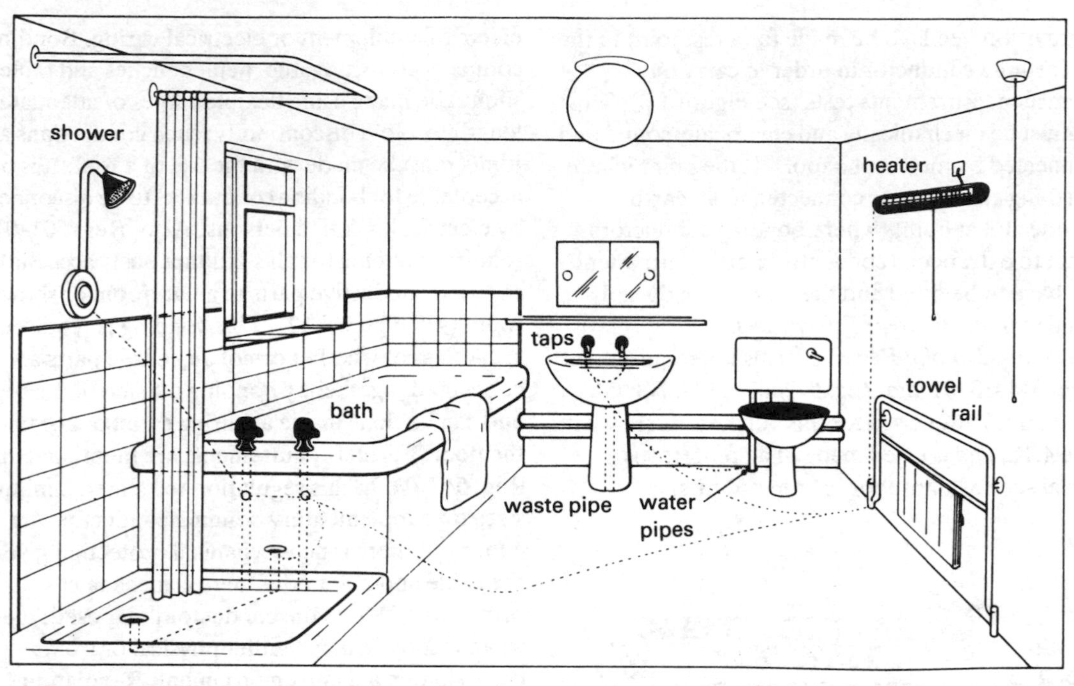

Figure 4.12 *Supplementary bonding in a bathroom (shown by dotted line)*

although they are not required for Class II equipment, since such equipment relies on double-insulation for safety, the circuits themselves need to have them incorporated. Their sizes are to be determined by the regulations mentioned above. Ring final circuit protective conductors must be run in the form of a ring similar to the phase and neutral conductors, and also bare circuit protective conductors are to be identified with the colours green/yellow according to Reg. 514–03–01. Flexible or pliable conduit cannot be used as a protective conductor but other metallic wiring systems can, such as metal conduit and trunking, cable armouring, etc., provided that they satisfy the two main regulations (Regs 543–02–04 and 543–02–05).

In terms of earth electrodes, Reg. 543–02–01 recognises that several different types can be used; some of these are shown in Figure 4.13. Often the choice will depend on the circumstances and the environment, such as the possibility of corrosion increasing the earth resistance (Reg. 542–02–03) or sometimes the effects of soil drying out or actually freezing. The materials used for earth electrodes should be capable of withstanding damage due to corrosion. In some cases more than one earth electrode may need to be installed to obtain a low enough value and the practice here is to allow a greater distance between the placing of the rods than their actual length. Table 54A of the IEE Wiring Regulations indicates the minimum cross-sectional areas allowed for buried earthing conductors when they are: (a) protected against corrosion, and (b) when they are not protected against corrosion, and depending on whether or not they are mechanically protected. Figure 4.14 shows one such arrangement of terminating an earth electrode and giving it full protection, as well as the facility for inspection. It is important that the earth electrode is made accessible and that it is also properly labelled in accordance with Reg. 514–13–01. Also of importance is the mechanical protection required for the earthing conductor as it emerges out of the soil and into the consumer's premises: this is often achieved with a short piece of conduit or robust PVC pipe.

It should be pointed out again that earthed equipotential bonding and automatic disconnection

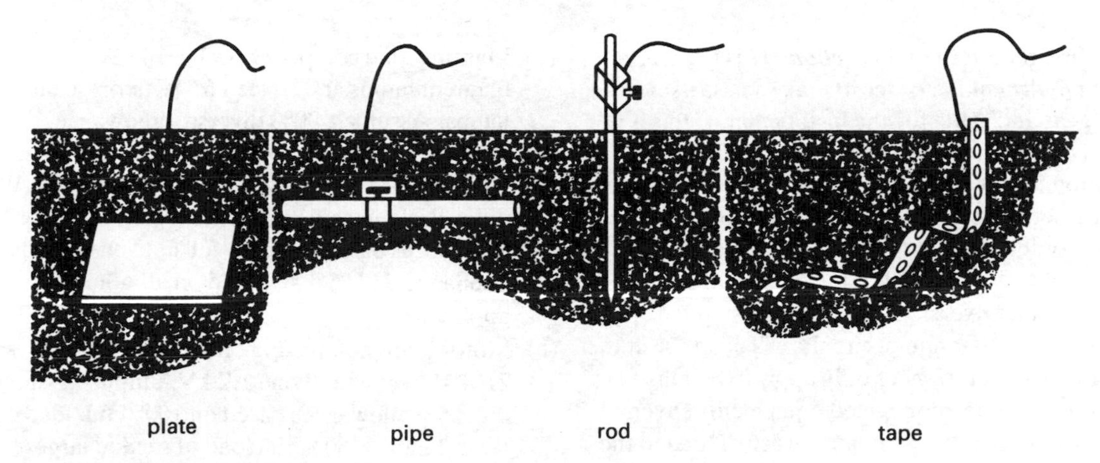

Figure 4.13 *Types of earth electrode*

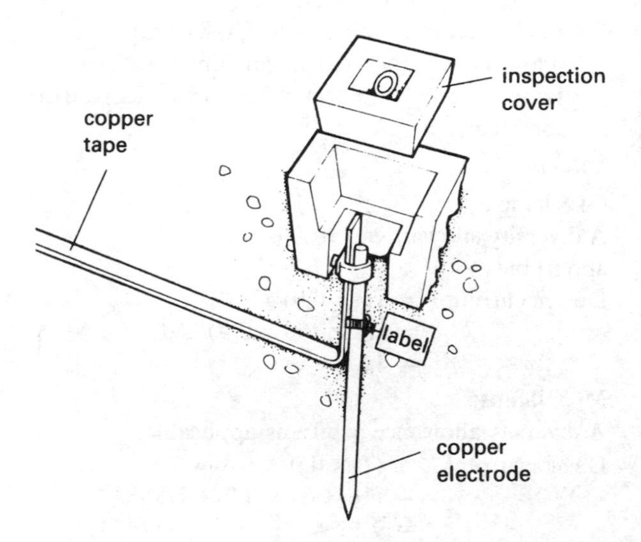

Figure 4.14 *Earth electrode termination*

of supply is the most important method of protection against indirect contact. For other methods, see Reg. 413–01–01.

Installation design

The assessment of general characteristics in Part 3 of the IEE Wiring Regulations covers the salient points in the design of an electrical installation. Reference to this part will show that there are a number of important areas to be considered, such as: its intended purpose; general structure and supply; the effects of external influences; compatibility and maintainability of installed equipment.

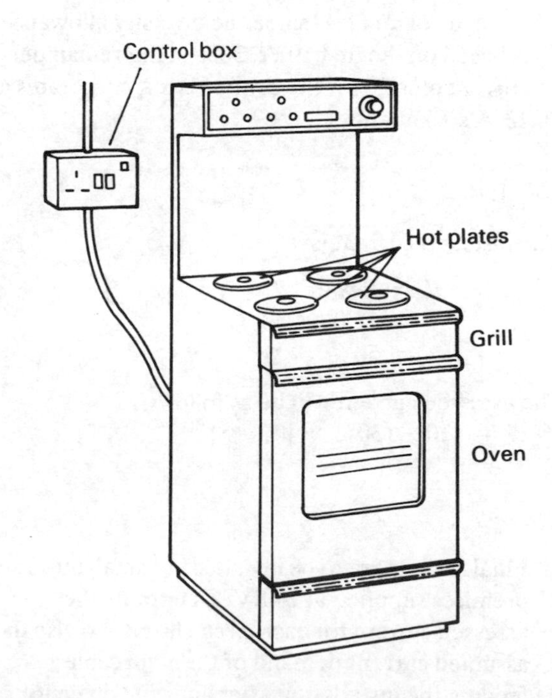

Figure 4.15 *Cooking appliance*

The first requirement in *Section 311* is to ascertain *maximum demand* in order to make an assessment of switchgear and cable for the installation. If this is not carried out, then the installation may be uneconomic and plant and equipment may never be fully utilised. The aim, therefore, is to apply diversity, which should cater for the present needs of the building as well as provide a facility for future development. Diversity allowances are applied on the understanding that not all circuits will be in use at the same time and allowances only apply to certain specific circuits and connected equipment. Diversity allowance is the ratio of minimum actual load to the installed load.

The designer of the electrical installation will have all the necessary information at hand concerning the whereabouts of equipment and distribution boards, and the following examples will serve to illustrate how diversity is applied.

Example

1 What is the assessed current demand for a household cooking appliance rated at 12 kW, 240 V having a 13 A socket outlet incorporated in the control unit? Assume the diversity allowance is based on the first 10 A, 30% of the remainder current plus 5 A if the control box incorporates a 13 A socket outlet.

Solution

The installed current is

$$I = \text{Power } (P) \div \text{Voltage } (V) = 12\,000/240 = 50 \text{ A}$$

The assessed current will be as follows:

$$I = [10 + (30\% \times 40) + 5] = 27 \text{ A}$$

Example

2 Final circuits are to be installed in a small business premises supplied at 240 V. Determine the assessed current for each circuit listed and also the assumed current demand of the main cable feeding the installation after applying diversity allowance.

(a) Lighting load comprises twelve 100 W filament lamps and sixteen 65 W fluorescent lamps. Assume a 90% diversity allowance.
(b) Heating and small power load comprising one 3 kW heater, two 1 kW heaters and two 500 W office desk machines. Assume diversity allowances based on 100% full load on the largest appliance and 75% full load on all other appliances.
(c) Motor loads comprising one 4.8 kVA input, one 2.4 kVA input and one 1.2 kVA input. Assume diversity allowances based on 100% full load of largest motor, 80% full load of second largest motor and 60% full load of third motor.
(d) Water heater load of 5 kW. No diversity allowance applicable. Socket outlets to BS 1363 comprising one radial circuit (32 A) and one radial circuit (20 A). Assume a diversity allowance of 100% of current demand of the largest circuit and 50% for the current demand of the second circuit.

Solution

GLS lamps:
A diversity allowance of 90% is applicable

$$\text{Design current } I = (P \times 0.9)/V = (12 \times 100 \times 0.9)/240 = \mathbf{4.5\ A}$$

MCF lamps:*
A diversity allowance of 90% is applicable

$$\text{Design current } I = (P \times 0.9 \times 1.8)/V = (16 \times 65 \times 0.9 \times 1.8)/240 = \mathbf{7.02\ A}$$

Heating and power:
A diversity allowance of 100% f.l. of largest appliance plus 75% f.l. of remaining appliances

$$\text{Design current } I = [(P \times 1)/V + (P \times 0.75)/V] = [3000/240 + (3000 \times 0.75)/240] = 12.5 + 9.375 = \mathbf{21.875\ A}$$

*For a discharge lamp, without information about control gear losses and power factor, a 1.8 multiplier is used with the lamp wattages.

Motors:
A diversity allownace of 100% f.l. of largest motor plus 80% f.l. of second largest motor plus 60% f.l. of remaining motors.

Design current $I = [(P \times 1)/V + (P \times 0.8)/V + (P \times 0.6)/V$
$= [4800/240 + (2400 \times 0.8)/240 + (1200 \times 0.6)/240]$
$= 20\text{ A} + 8\text{ A} + 3\text{ A}$
$= \mathbf{31\text{ A}}$

Water heaters:
There is no diversity allowance.

Design current $I = P/V$
$= 5000/240 = \mathbf{20.83\text{ A}}$

Socket outlets:
A diversity of 100% of current demand of largest circuit plus 50% of current demand of every other circuit.

Design current $I = 32 + (0.5 \times 20)$
$= \mathbf{42\text{ A}}$

The total assumed current demand is as follows:

$I = 4.5 + 7.02 + 21.875 + 31 + 20.38 + 42$
$= \mathbf{127\text{ A}}$ (approx.)

Figure 4.16 shows the electrical services of a typical office building. For a three-phase, four-wire system, the installation designer tries to balance the loads over the three phases of the incoming supply so that the cables and switchgear can be economically chosen. Section 313 of the IEE Wiring Regulations requires the designer to consider the following characteristics.

(a) nominal voltage(s),
(b) nature of the current frequency,
(c) prospective fault current at the origin of the installation,
(d) earth fault loop impedance of that part of the system external to the installation,
(e) suitability for the requirements of the installation, including the maximum demand,
(f) type and rating of the overcurrent protective device acting at the origin of the installation.

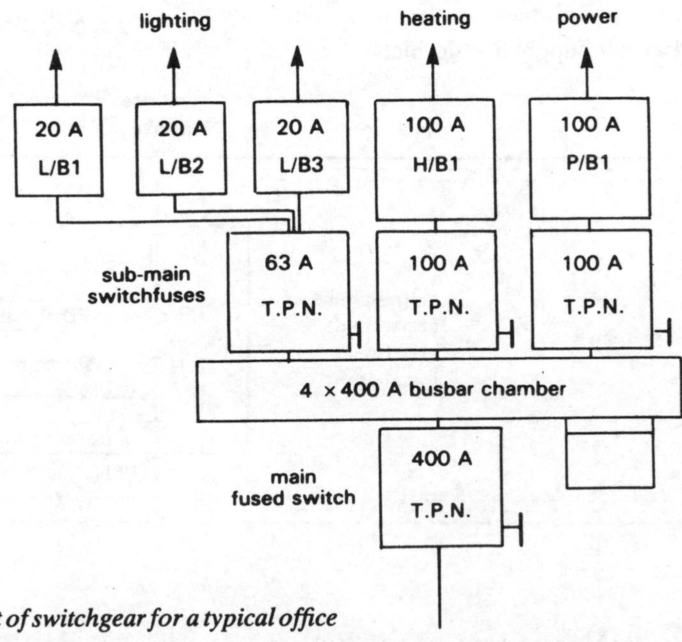

Figure 4.16 *Layout of switchgear for a typical office building*

In practice, the designer will obtain a lot of this information from the local REC. A domestic consumer will be given a supply of 240 V a.c. at a frequency of 50 Hz and in most cases the service arrangements will allow for a maximum current of 100 A and, to satisfy (c) above, the REC will protect the installation by fitting a BS 1361 type I high-breaking-capacity cartridge fuse at the intake position. The reason for using this type of fuse is that, depending on where the supply transformer is sited in the system, consideration has to be given to the level of fault current that is likely to flow and cause damage. The closer a consumer is to the supply transformer the greater will be the magnitude of any possible fault. The table shows the estimated maximum prospective fault currents likely to occur in different lengths of service line using either a 16 mm² copper or 25 mm² aluminium service cable.

Length of service line (m)	*Prospective current* (kA)
0	16.0
2	13.7
4	11.7
6	10.1
8	8.8
10	7.8
12	7.0
14	6.3
16	5.7
18	5.3
20	4.9
30	3.5
40	2.7
50	2.2

From the point of view of the installation's *external earth fault loop impedance* (Z_E), typical maximum values for the three types of earthing system previously mentioned are: (a) TN–C–S system, 0.35Ω; (b) TN–S system, 0.8Ω; and (c) TT system, 21Ω.

The external earth fault loop impedance is that part of the distribution system on the REC's side of the cut-out fuse, and it will comprise the route taken by a fault current as it flows between the phase conductor and PEN conductor via the main supply transformer.

Figure 4.17 shows a diagram of this impedance path as well as the total *earth fault loop impedance path* (Z_s) which is made up of Z_E and the *resistances* of both the *phase conductor* (R_1) and *protective conductor* (R_2).

The following formula is used to determine by calculation the actual earth fault impedance: $Z_s = Z_E + R_1 + R_2$.

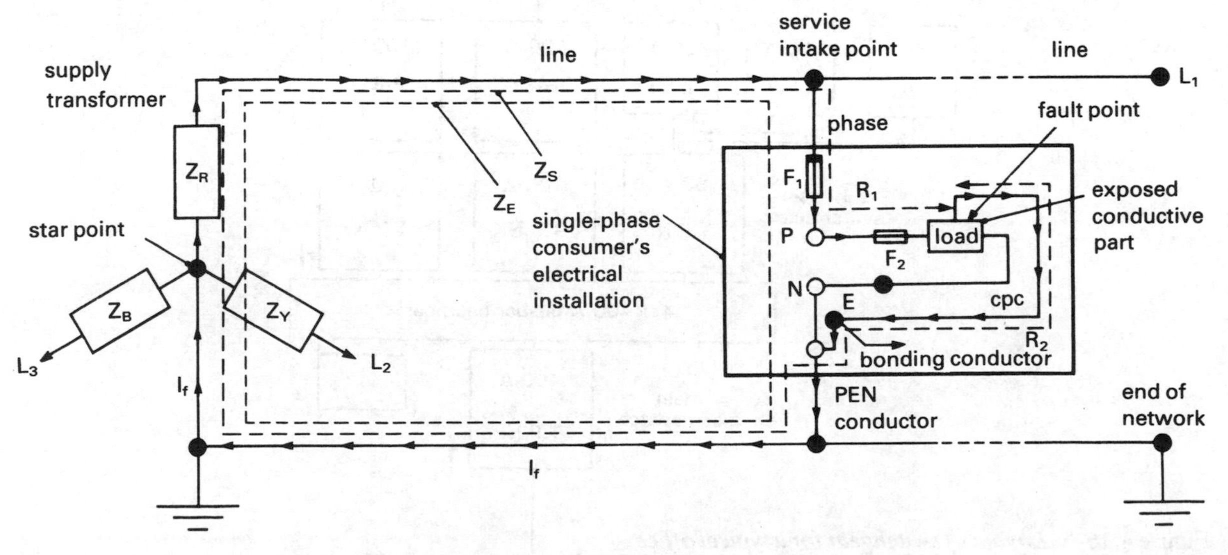

Figure 4.17 *Earth fault loop impedance path for TN–C–S system*

For the sake of clarity, only one final circuit has been included in the consumer's wiring and the total earth fault loop impedance path of this circuit is the route taken by the earth fault current (I_f). It will be seen that if this occurred at the load, the leakage current would pass into the circuit protective conductor (c.p.c.) and then flow into the PEN conductor to the star point of the supply transformer. It would then flow back to the point of fault via the transformer's red phase winding, distributor's line and final circuit. The object is to cause automatic disconnection of the supply and since the final circuit fuse will be of a lower rating than the cut-out fuse, it should disconnect this part of the circuit in the required times of 0.4 or 5 s as mentioned earlier. It should be noted from this type of earthing system that if a short circuit occurred between live conductors (e.g. phase and neutral), it would become an earth fault at the junction where the PEN conductor separates. It is most important to remember that the TN–C–S system reverts to a TN–S system after the energy meter and it is quite in order to install a residual current device to detect earth leakage current. This means of course that if such a device was installed it would disconnect the whole system before any circuit protective device operated, that is of course if it was connected as a main switch controlling all final circuits.

Circuit protective devices

The fundamental safety requirement concerning these devices is given in Reg. 13–03–01 of the IEE Wiring Regulations which states that every installation and every circuit shall be protected against overcurrent. A protective device must operate for two *overcurrent* conditions, namely, *overload* and *short circuit*. A meaning of these two terms can be found in the Terminology chapter of this book.

Chapter 43 and Section 473 of the IEE Wiring Regulations are both concerned with the requirements for protection against overcurrent. Reg. 434–02–01 requires the prospective fault current, under short circuit or earth fault, to be determined at every relevant point of the complete installation. This is to safeguard those parts of the installation (switchgear, cables, joints, terminations, etc.) against damage whenever faults of negligible impedance occur. Fault levels are much higher the closer such equipment is placed to the origin of the installation and they can produce considerable thermal and electromagnetic stress on circuit conductors (see Figure 4.18).

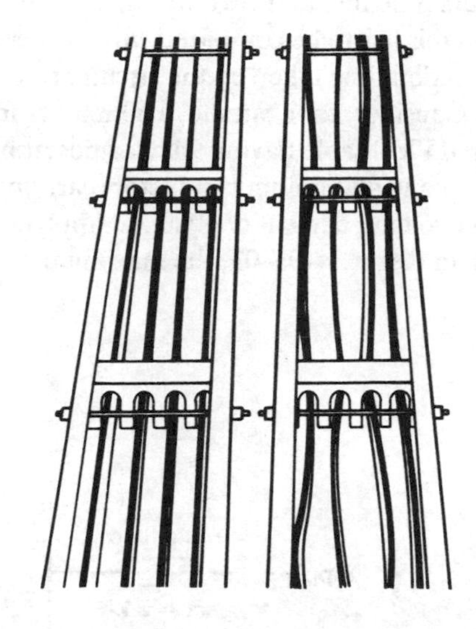

Figure 4.18 *Results of a short circuit in a rising main trunking system*

Reg. 434–03–01 allows the use of a protective device having a breaking capacity less than the prospective fault current at the point where it is installed, provided there is a device on the supply side which has the necessary breaking capacity. The characteristics of both devices must be co-ordinated. Reg. 434–03–02 allows one device to be used for both overcurrent conditions provided its breaking capacity is not less than the prospective fault current at its point in the installation.

Consideration has to be given to such factors as fault clearance time, conductors connected in parallel and certain types of circuit breaker. Reg. 434–03–02 provides a formula for finding the time in which a given fault current will raise the conductor temperature to its limiting value. This formula is known as the *adiabatic equation* and refers to a condition which is without gain or loss of heat. This means that when a cable conductor carries current and becomes heated under fault circuit conditions, the temperature of it rises very quickly. The short circuit will persist for a short time (often less than 5 s) and measures need to be taken to see that the cable insulation is not damaged by the let-through energy released. The duration time (t) in the equation is the maximum time the fault current can be allowed to flow and this has to be compared with the actual disconnection time taken by the circuit protective device. Consider, for example, a 50 mm² twin armoured PVC cable having 90° thermosetting insulation and aluminium conductors carrying a prospective fault current of 4000 A. Applying the formula in Reg. 434–03–03, the maximum disconnection time for the circuit protective device is found to be:

$$t = k^2S^2/I^2$$
$$= 94 \times 94 \times 50 \times 50/ 4000 \times 4000$$
$$= 1.38 \text{ s.}$$

See Table 43A, IEE Wiring Regulations.

For very short durations of less than 0.1 s, the above formula can be rearranged to become the '*let-through energy*', and the characteristics of a protective device can be obtained from manufacturers' catalogues.

Further requirements for protection against overcurrent are to be found in Reg. 473–01 which refers to the position of the devices for overload protection, i.e. at a point where a reduction occurs in the current-carrying capacity of conductors. An

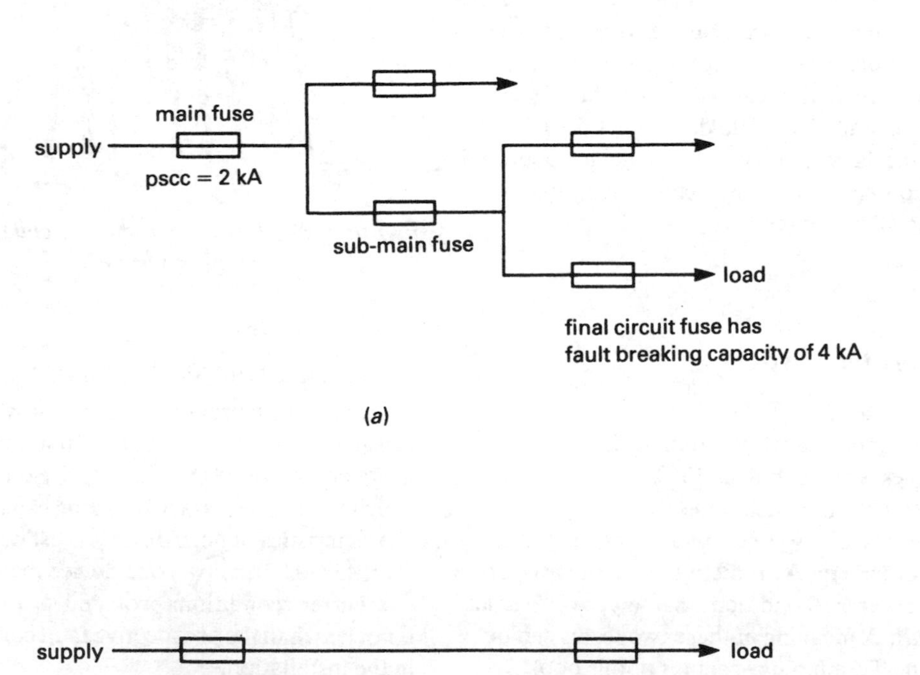

Figure 4.19 *Circuit protection satisfying Reg. 434–03–01*

Figure 4.20 *Types of circuit protective devices*

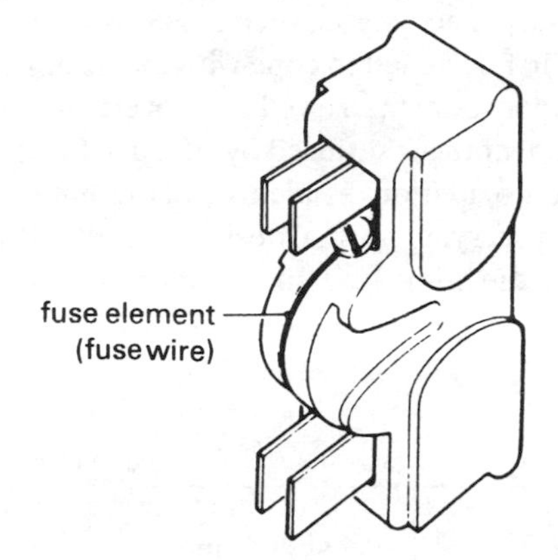

(a) BS 3036 rewirable fuse

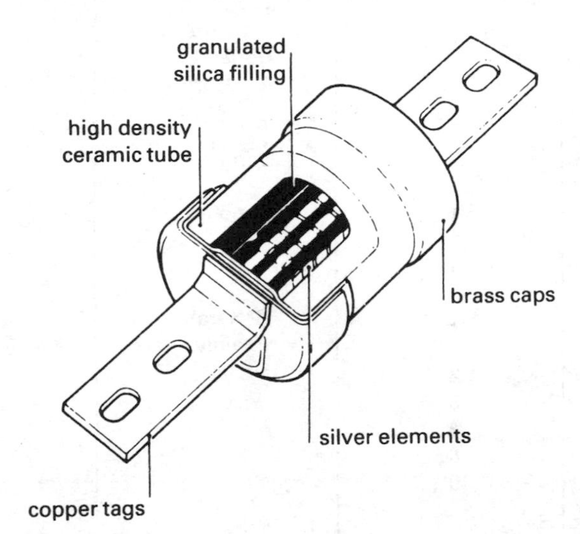

(b) BS 88 Part 2 h.b.c cartridge fuse

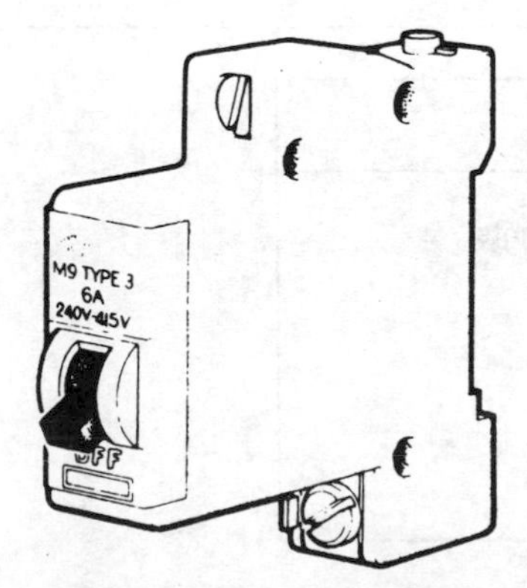

(c) BS 3871 miniature circuit breaker

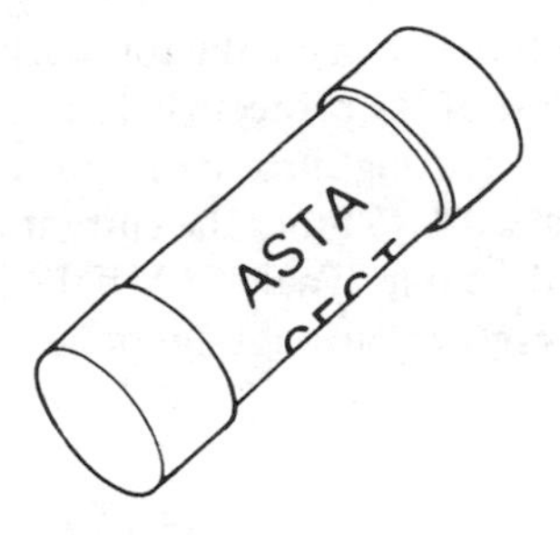

(d) BS 1361 cartridge fuse

example of this is in a ring final circuit where a reduced conductor size is taken from a fused spur outlet. Reg. 473–01–02 allows a protective device to be placed along the circuit route provided that the part of the run between the point where the value of current-carrying capacity is reduced and the position of the protective device has no branch circuits. There are situations where the protective device can be omitted (Reg. 473–01–03), as for example in the case of the secondary circuit of a current transformer which could create a large and dangerous induced e.m.f. if the protective device operated and opened the circuit. This could also apply to exciter circuits and electromagnet circuits. Another example where overload protection is not necessary is in a fixed heating load which cannot become overloaded by design.

There are also certain conditions imposed on the positioning of protective devices, such as at a point of change in current-carrying capacity, where the distance from the point of change must not exceed 3 m and where there is no risk of short circuit, fire or danger to persons. Figure 4.20 shows several common protective devices. The *BS 3036 semi-enclosed fuse* has a number of well-known disadvantages such as deterioration through age, the fact that the wrong size fuse element can be fitted, poor discrimation with other protective devices and relatively low breaking capacity. It also has a relatively high fusing factor (i.e. the ratio of rated minimum fusing current and current rating) approaching 2:1 which incurs a correction factor of 0.725 and this could mean installing a larger cable than would otherwise be necessary. An explanation of the 0.725 factor is found in Appendix

4(5) of the IEE Wiring Regulations which recognises the temperature rise in conductors during the operating time of the protective device. For safety reasons the operating current of a protective device must not exceed 1.45 times the current-carrying capacity of the cable. Table 53A of the IEE Wiring Regulations shows various sizes of fuse element available.

Cartridge fuses, particularly those described as the high breaking capacity type, such as *BS 88 Part II fuses* were first designed to cope with increasing fault levels on supply systems as well as to overcome the serious disadvantages suffered by BS 3036 fuses in terms of their oxidation problems and premature failure when carrying normal design currents. It will be seen from the h.b.c. fuse link in Figure 4.20(b) that

Protective device	Cat. duty	Rated breaking capacity (kA)	Application
B.S. 88 Cartridge fuses		80	Industrial & General
B.S. 1361 Cartridge fuses		16.5 33	Domestic & General
B.S. 1362 Cartridge fuses		6	Domestic
B.S. 3036 Semi-enclosed rewireable fuses	S1 S2 S4	1 2 4	General " "
B.S. 3871 (Part 1) Miniature circuit breakers	M1 M1.5 M2 M3 M4 M6 M9	1 1.5 2 3 4 6 9	General (mainly final circuits) " " " " "
B.S. 4752 (Part 1) moulded case circuit breakers		According to design	Industrial

B.S. 3871 m.c.b. characteristics

Type	Instantaneous tripping currents
1	2.7 I_n to 4.0 I_n
2	4.0 I_n to 7.0 I_n
3	7.0 I_n to 10.0 I_n
4	10.0 I_n to 50.0 I_n

Figure 4.21 *Common circuit protective devices*

it comprises silver fusible elements and these are contained in rigid tubes filled with an arc-quenching powder of granulated silica or graded quartz. This is done to prevent the formation of an arc and allows them to quickly clear short circuit faults. This design gives them a high degree of discrimination with other protective devices, and they also have non-ageing characteristics and are insensitive to changes in ambient temperature. Another extremely good and inexpensive cartridge fuse is the *BS 1361 Type 1 fuse* and from Figure 4.21 it can be seen that it is quite capable of handling fault levels of 16 kA as previously mentioned. In the smaller sizes, these fuses are still ideal overcurrent protection devices and are an economical alternative to miniature circuit breakers and the rewireable semi-enclosed fuses. They also provide ideal fuse discrimination at very fast disconnection times as will be seen in the time–current characteristics of Figure 4.22. From the BS 1361 fuse. graph, it will be seen that for a 5 second disconnection time, the 5 A device will need to take a current of between 10A and 20A. The families of curves that are produced start with the lowest fuse rating first. Both axes are scaled logarithmically which means that each marked graduation point changes by a factor of ten. To check your answer, look at Table 41D of the Wiring Regulations and divide the nominal supply voltage of 240 V by the stated maximum earth fault loop impedance of 17.1Ω. This should give a rupturing current of 14 A approximately.

BS 3871 Part 1, miniature circuit breakers (m.c.b.s) are another important circuit protective device giving protection against overcurrent and fire risk. They are primarily used in final circuits and are classified according to their instantaneous tripping current, i.e. the current at which they will operate within 100 ms. Figure 4.21 shows four types of these devices together with their respective bands of magnetic operation. The Type 1 trips between 2.7 and 4 times its rated current and is suitable for loads with little or no switching surges such as in domestic consumer application. The Type 2 trips between 4 and 7 times its rated current and has a more general use in that it provides a lower chance of nuisance tripping combined with a slower operation on heavier loads. The Type 3 will be seen to have an instantaneous trip between 7 and 10 times its rated current and is suitable for industrial use such as in motor circuits, lighting banks and even transformer circuits. Type 4 will be seen to have a tripping current in excess of 10 times its rated current and it can be used in welding circuits and even medical X-ray equipment. It will be noticed from the inverse time–current characteristic of this device that the operating curve shows two distinct shapes. The upper curve relates to the thermal operating performance during an overload whereas the lower curve relates to its electromagnetic performance during a short circuit. The thermal operation will cause deflection of a bimetal heater and the greater the overload becomes the faster the m.c.b. operates. This performance is followed by a more dramatic operation when a short circuit occurs but it must be remembered that the above types have built-in time delays. Figure 4.23 shows a simple circuit diagram of an m.c.b. with its two essential operating mechanisms clearly indicated. On short circuit, the in-rush of current through the device will be excessive and the armature of the electromagnet will rapidly operate against the moving contact mechanism; the latch and spring arrangements work in conjunction with each other causing the moving contact to separate very quickly. The arc so formed will be extinguished in a separate chamber which comprises a number of baffles or arc chutes.

Another similar type of circuit breaker is the *BS 4752 Part 1 moulded case circuit breaker (m.c.c.b.)*. These have ratings ranging from about 30 A to 800 A and are capable of operating at higher fault levels. They have wide industrial application.

Where more than one protective device is fitted in a circuit, Reg. 533–01–06 points out that to prevent danger the characteristics and setting shall be such that proper discrimination is achieved (see Figure 4.24). Discrimination of operation between protective devices is best understood by studying fuse manufacturers' time–current characteristics. Basically, the device nearest the fault should operate first, leaving other devices intact without loss of supply to the circuits they protect. Also worth mentioning is back-up protection which is used where fault levels are very high. Types of protective device to meet this need are either BS 88 Part 2 cartridge fuses or BS 1361 cartridge fuses. There are special fuses available for motors circuits which tolerate large in-rush currents.

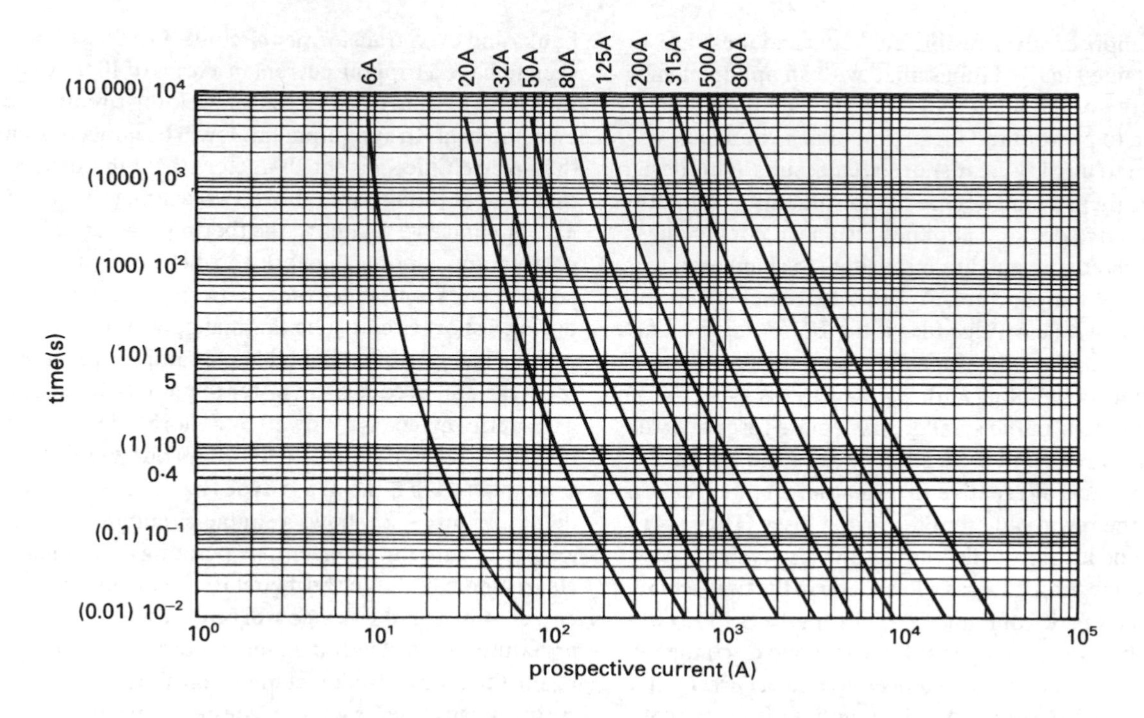

(a) BS 88 Part 2 fuse characteristics

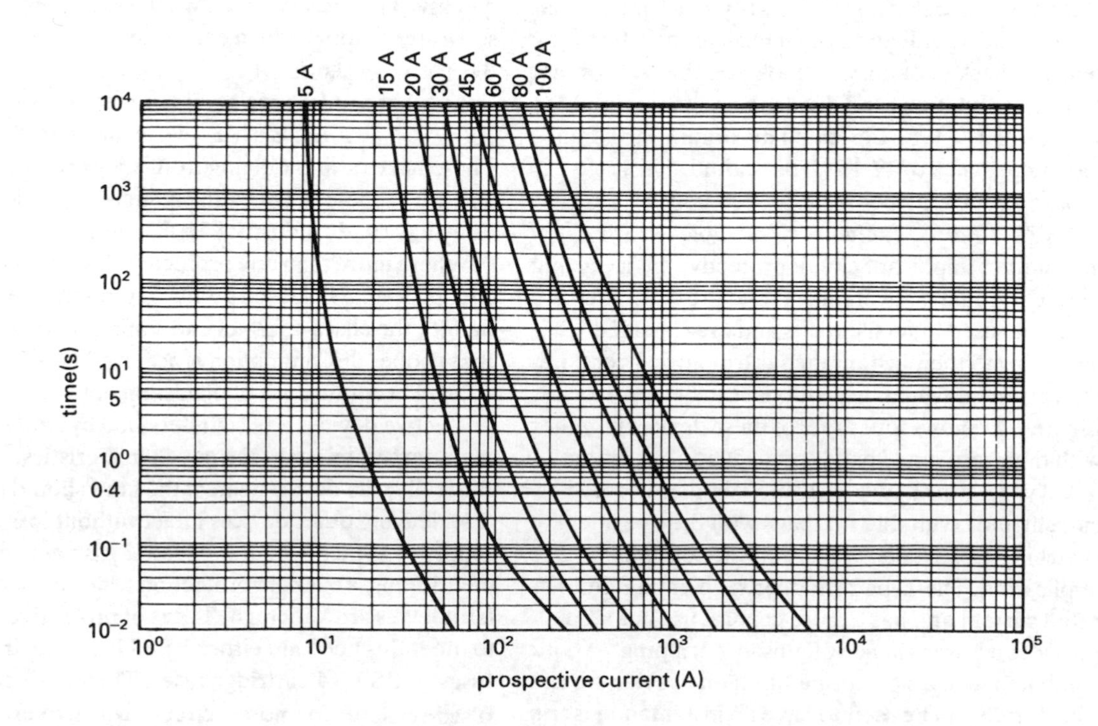

(b) BS 1361 fuse characteristics

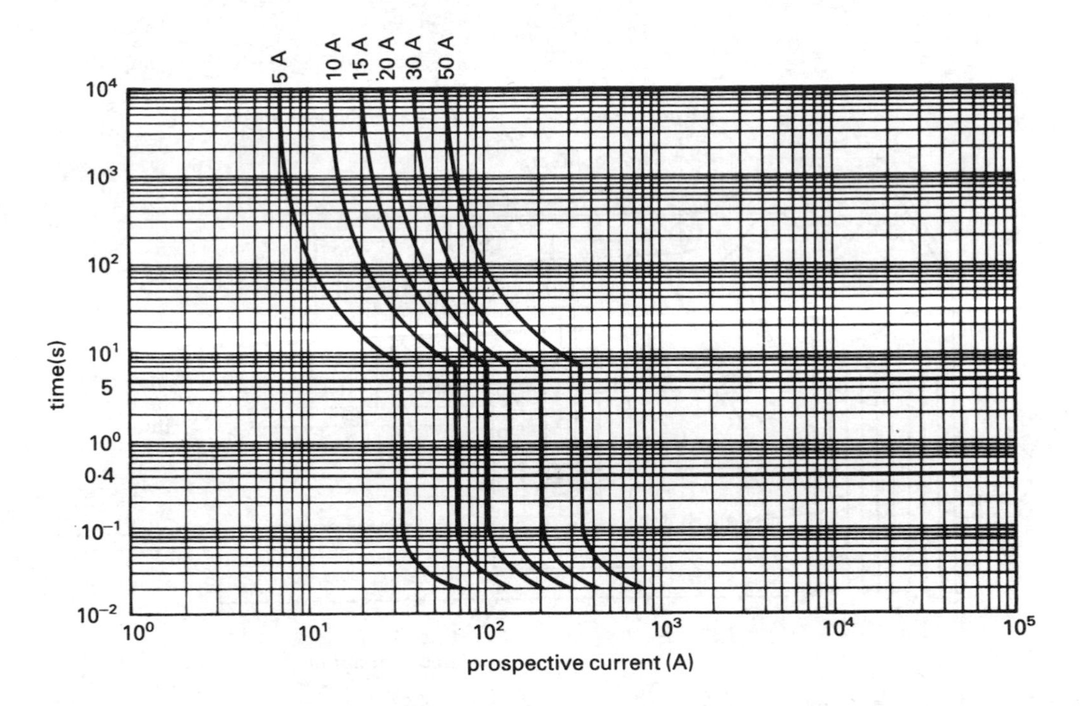

(c) BS 3871 Type 2 miniature circuit breaker characteristics

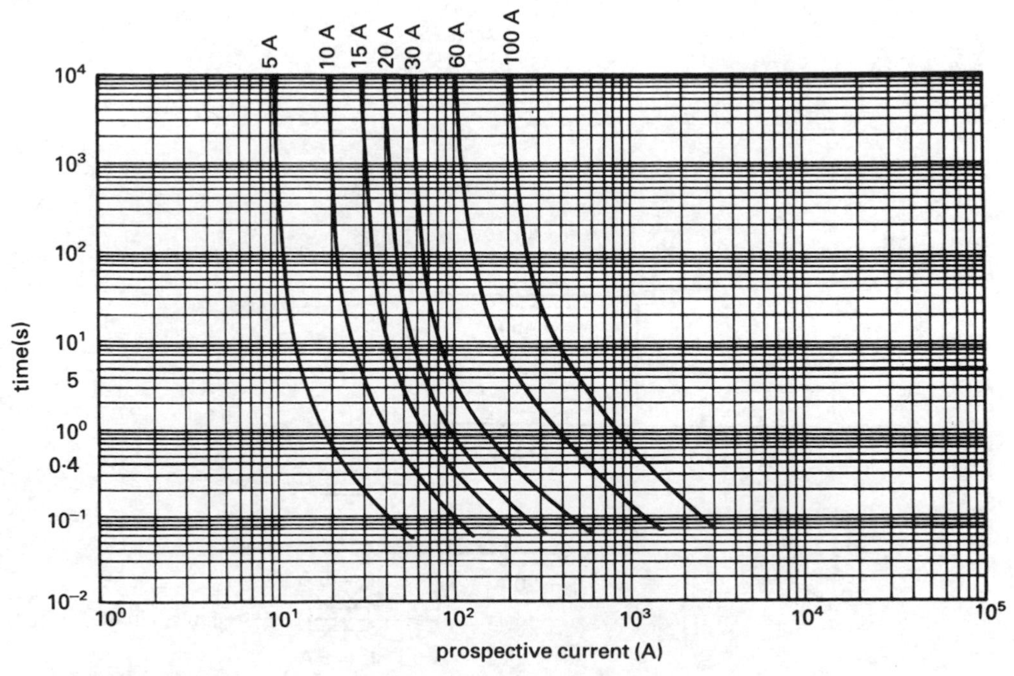

(d) BS 3036 fuse characteristics

Figure 4.22 *Time-current characteristics of protective devices*

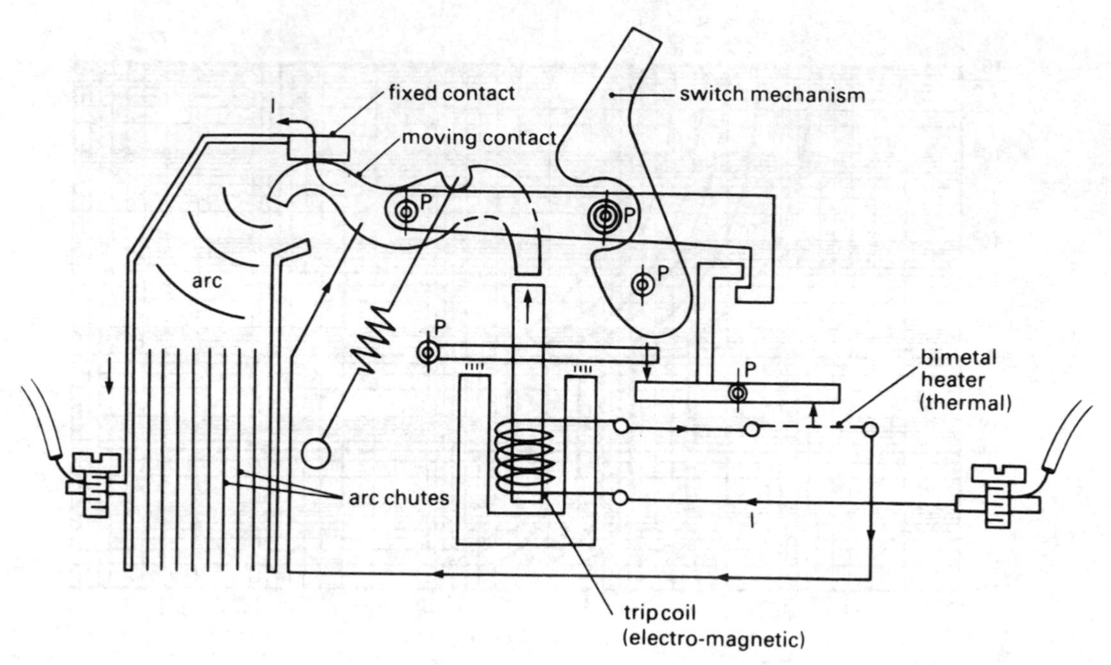

Figure 4.23 *Circuit diagram of m.c.b.*

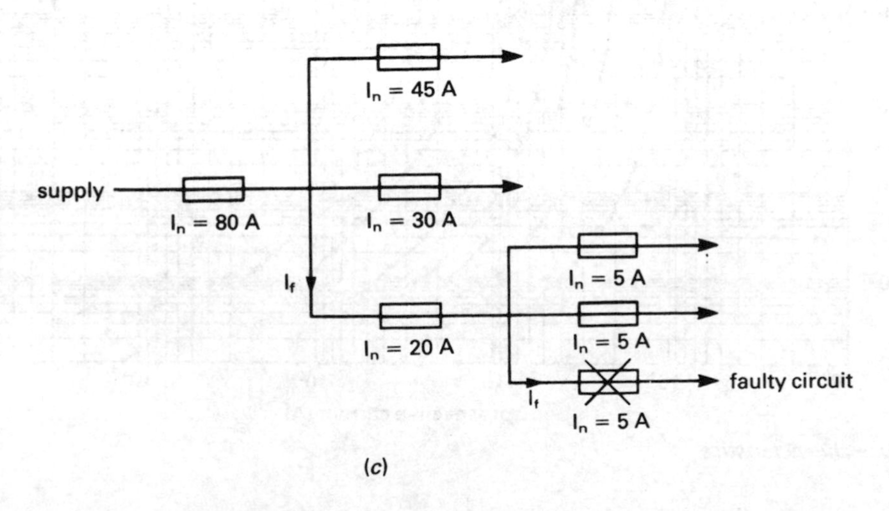

Figure 4.24 *Fuse discrimination satisfying Reg. 533–01–06*

Cable selection procedures

The designer of an installation has to know the 'conditions within' an installation before he can begin to select and size the cables for circuits. Wiring systems are discussed in the next chapter but the designer will need to consider any adverse conditions that may affect the wiring if it is not adequately constructed and protected; for example, corrosion of metallic enclosures, extremes of temperature, effects of the weather, or a dirty, dusty, flammable or explosive atmosphere. Appendix 5 of the IEE Wiring Regulations provides a classification of external influences based on the environment, utilisation and building construction.

With regard to the sizing of conductors and cables for circuits, Appendix 4 consists of eight subheadings to give the designer guidance. Table 4A provides a schedule of the methods of installing cables, Tables 4B1 and 4B2 concern correction factors for cables in groups and Table 4B3 is concerned with cables installed in enclosed trenches. Table 4C1 and 4C2 deal with correction factors for ambient temperature and numerous tables are to be found for different types of cable, different types of insulation and different types of sheath. The tables consider cables that are made of copper and aluminium.

The following are some of the important factors which need to be discussed before any examples are given:

(a) design current and rating of protective device(s),
(b) control of voltage drop,
(c) circuit disconnection and shock protection,
(d) thermal constraints of circuit conductors.

The design current (I_b) can be regarded as the load current of a circuit. It is the first calculation to make and is often found from the formula: $I = P/(U \times p.f.)$ where P is the power rating of the load in watts (W), U is the supply voltage in volts (V) and *p.f.* is the power factor of the circuit. The current setting (I_n) of the protective device for the circuit is, under Reg. 433–02–01, allowed to be equal to but not less than I_b. Typical values are to be found in Tables such as 41B1 and 41C. Voltage drop must not affect the safe functioning of electrical equipment in normal service and Reg. 525–01 limits it to 4% in the final circuit where a relevant British Standard is not applicable.

Reg. 531–01 refers to overcurrent protective devices which are used for protection against electric shock. Their operating time not only has to be appropriate to the magnitude of fault current so that the permissible final temperature of the protective conductor is not exceeded but also has to be appropriate to the disconnection times stated in various regulations of Section 413, namely 0.4 s and 5 s.

In a TT earthing system the earth fault loop impedance may be too high and shock protection by indirect contact better served with the use of a residual current device, particularly for socket outlet circuits which are a requirement of Reg. 471–08–06. Further requirements for r.c.d.s in TT systems can be found in Regs 413–02–18 to 413–02–20 as well as in Section 531–02.

The requirements for thermal constraints are contained in Section 434 and have already been dealt with under circuit protective devices. Reg. 434–03–03 is applicable and gives a formula for determining whether a short circuit current occurring at any point in a circuit will be interrupted in a time which will not allow the cable conductors to exceed their admissible limiting temperature. This is very important because the heat produced by a short circuit is contained within the cable conductor cores. The formula is rearranged in Reg. 543–01–03 in order to determine by calculation the minimum acceptable size of protective conductor.

Whilst Appendix 4(6) provides the procedure, along with supporting regulations and tables, for the selection of an appropriate cable size, you might like to consider the following as a step-by-step approach.

Step 1 Find the design current (I_b) and select *an overcurrent protective device* (I_n) from the tables related to the earthing system or disconnection time required.

Step 2 Find values of any correction factor (C) used in your circuit 'conditions within', e.g. for ambient temperature (C_a) see Table 4C1 where the protective device intends to provide protection against short circuit; otherwise use Table 4C2 for overload protection where a BS 3036 fuse is used; for grouping (C_g) see Tables 4B1, 4B2 or 4B3; for thermal insulation (C_i) see Reg. 523–04–01 and information in Table 4A. A correction factor of 0.725 is required for a BS 3036 fuse.

Step 3 Find the effective current-carrying capacity of the cable (I_z) by using the formula $I_z = I_n/C$.

Step 4 Consider the wiring methods in Table 4A and select from the appropriate tables listed (see Tables 4D1A to 4L4A) a suitable tabulated current (I_t) of a given cable such that it is equal to or greater than I_z.

Step 5 From the tables of voltage drop, show that the selected cable's voltage drop does not exceed either a stated value or that given by Reg. 525–01. Here $V \leqslant 4\%$ of U_0 which is checked by the formula $V = L \times I_b \times$ mV/A/m.

Step 6 To satisfy circuit disconnection time (shock protection), see the appropriate values stated in various tables.

Step 7 To satisfy the thermal constraints of a chosen protective conductor (S) use the formula in Reg. 543–01–03. Alternatively, the protective conductor size can be determined from Table 54G.

Consider the following examples:

Example 1

An electric heater rated at 3 kW/240 V is to be installed as a fixed appliance. The feed will be from an existing spareway in a fuseboard containing a BS 88 Part 2 fuse. A twin-with-earth PVC-insulated and sheathed cable (copper conductor) will be run with three similar cables in an ambient temperature of 25°C, bunched and clipped direct and totally surrounded by thermal insulation for most of its 12 m run. The voltage drop of the circuit must not exceed 6 V.

(a) Determine the design current (I_b) and the normal rating of the fuse (I_n).
(b) Calculate the minimum current rating of the required cable (I_z).
(c) State the table and columns used in the IEE Regulations to determine the initial size of the cable.
(d) Select the minimum size of cable to comply with the requirements for current-carrying capacity and voltage drop.
(e) Calculate the actual voltage drop.

Solution

The procedure follows the steps mentioned above.

(a) $I_b = P/U_o = 3000/240 = 12.5$ A.
From Table 41D, I_n is chosen to be 16 A.
(b) From Table 4C1, $C_a = 1.03$ and from Table 4B1, $C_g = 0.65$.
From Reg. 523–04–01, $C_i = 0.5$.
Thus $I_z = I_n/(C_a \times C_g \times C_i) = 16/(1.03 \times 0.65 \times 0.5) = 47.8$ A.
(c) Table 4D2A, columns 1 and 6.
(d) A 10 mm^2 is chosen having a current rating of 63 A and voltage drop of 4.4 m V/A/m.
(e) $V = L \times I_b \times$ mV/A/m $= 12 \times 12.5 \times 0.0044 = 0.66$ V.

Example 2

A 12 kW/240 V cooking appliance is to be installed in a domestic dwelling connected to a TN–C–S earthing distribution system, where Z_E is to be regarded as 0.35Ω. If the cooker control point is 25 m away from the main consumer unit and it is wired in PVC/PVC/c.p.c. cable, select a suitable size cable for the final circuit. The following assumptions are made:

(a) the ambient temperature is 30°C,
(b) circuit protection is by means of a BS 1361 fuse,
(c) there is a thermal insulation factor to consider affecting one side of the cable,
(d) the cable is partly clipped throughout its run,
(e) the control point incorporates a 13 A socket outlet.

Solution

The procedure is as follows:

1 The *design current* (I_b) $= P/V$
$= 12\,000/240 = 50$ A

Applying diversity allowance then:
$I_b = [10 + (0.3 \times 40) + 5]$
$= 27$ A.

2 The *nominal setting* of the fuse (I_n) is 30 A. This is found from Table 41B1, IEE Regulations
$Z_s max = 1.2\Omega$.
3 See Regulation 523–04 and Table 4A for thermal insulation factor (Method 4).
4 Since there is only one correction factor, then:
$I_z \geqslant I_n/C_i$.
5 From table 4 D2A, columns 1 and 2, a 6 mm^2 cable is chosen, which has a *current-carrying capacity* (I_t) of 32 A and a voltage drop of 7.3 m V/A/m.
6 The voltage drop in this cable is found from:
$V = LIV/1000 = 25 \times 27 \times 0.0073$
$= 4.927$ V.

This is below the permissible value of 4% of 240 V, which is 9.6 V, and therefore the cable satisfies both the current requirement and the voltage drop.

Exercise 4

1 In the process of estimating the loading on the main supply cable for a hotel installation a diversity allowance is used on *five* cooking appliances. Determine the assessed current demand assuming *three* appliances are rated 15 kW/240 V and the remaining *two* are rated at 10 kW/240 V. Also assume that the diversity allowance is based on 100% for the largest appliance and 80% for the second-largest appliance and that the hotel is supplied with a three-phase, four-wire supply.

2 Explain the operation tests for a residual current device. State the meaning of the term 'sensitivity' with regard to its operation.

3 Show by means of a labelled diagram, how a domestic consumer's hot water storage cylinder and pipework are supplementary bonded to an electric immersion heater.

4 (a) Explain the term 'earth fault loop impedance path'.
(b) Using the formula in Reg. 543–01–03 of the IEE Wiring Regulations, determine the size of protective conductor suitable for use on a ring circuit operating at 240 V a.c. protected by a 30 A fuse to BS 3036. The circuit is wired in PVC-insulated cables (copper conductors) installed in PVC conduit. The value of earth loop impedance Z_s may be taken as 1.14Ω.

5 Explain what is meant by the following terms when applied to installation circuits:
(i) diversity
(ii) assumed current demand.

6 Using the BS 1361 fuse time–current characteristics (Figure 4.22(b)), estimate the disconnection times of the following fuses: 5 A, 15 A, 20 A and 30 A at a prospective current of 60 A. What do these times tell you?

7 State *two* advantages and *two* disadvantages of the following protective devices:
(i) BS 3036 semi-enclosed fuse
(ii) BS 88 Part 2 cartridge fuse
(iii) BS 3871 Type 1 miniature circuit breaker.

8 Explain with the aid of a diagram the term 'fuse discrimination' and state how it can best be achieved.

9 Describe how a miniature circuit breaker operates for
(i) an overload
(ii) a short circuit.

10 A 6 kW/240 V heating load is to be installed 10 m away from a distribution board using single-core PVC-insulated cables in plastic conduit. Circuit protection is by a BS 3036 fuse and the ambient temperature is 35°C. Select suitable size circuit conductors for the circuit. Assuming a 2.5 mm^2 cable is used as the protective conductor and that the value of Z_s is 0.977Ω, check that the thermal constraints of the protective conductor are satisfied using Reg. 543–01–03 of the IEE Wiring Regulations.

chapter five

Installation 2

After reading this chapter you should be able to:

1 state the advantages of a primary ring main and secondary rising main distribution system,
2 know the requirements for main switchboards,
3 state requirements for numerous wiring systems in common use today,
4 state a number of requirements for final circuits,
5 describe the operation and performance of several types of lamp, and draw diagrams of discharge lamp circuits,
6 know the requirements for emergency lighting and perform simple calculations on the lumen method for finding lamp flux,
7 state different factors affecting the choice of motors and know various circuit diagrams of motor starters,
8 know methods of fixing motors and carrying out maintenance procedures,
9 know the requirements for special installations, such as construction sites, farms and explosive installations.

Every electrical installation should be provided with the following means of control and protection.

(a) isolation
(b) overcurrent protection
(c) earth leakage protection.

When a high-voltage supply is required for a consumer, it may take the form of a *ring main*, i.e. a primary distribution system whereby the main cable is looped in the form of a ring. By doing this, any breakdown on the system can be isolated and repaired, while still maintaining supplies to the rest of the plant. Figure 5.1 is a line diagram of such an arrangement and it will be seen that the ring feeds four 11 kV/415 V step-down, delta-star transformers. These will be sited in separate local substations in order to feed low-voltage switchboards catering for secondary distribution requirements.

When the electrical installation requirements are needed to extend over a number of floors, it is usually the practice to provide a secondary distribution known as a *rising main*. This system often comprises

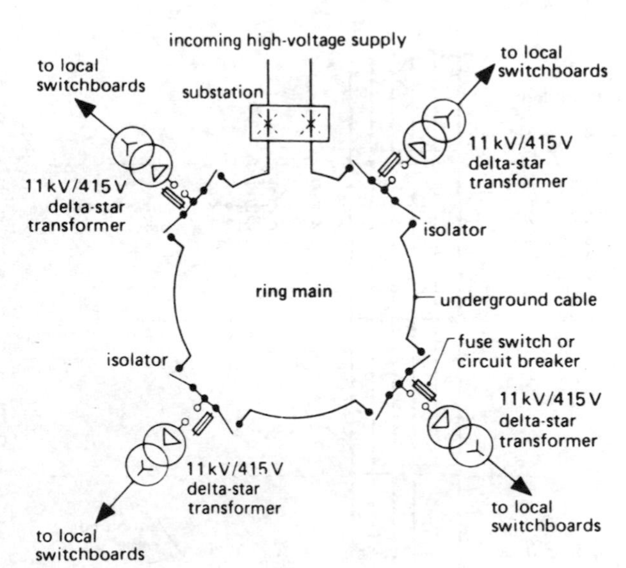

Figure 5.1 *Ring main primary supply distribution*

lightly insulated bare copper or aluminium busbars enclosed in metal trunking and suitable for loads in excess of 200 A per phase. It is usual to provide tap-off units at each floor level so that lateral supplies can be given. Numerous other wiring systems can be taken from it to feed final circuits on different floors. There are several IEE Wiring Regulation requirements which call for internal barriers to be provided in order to prevent air at the top of the system from reaching an excessively high temperature. In practice, one often finds that *fire barriers* are fitted between each floor or every 5 m. whichever is the lesser distance. They may also be required in other situations where cables or wiring systems pass through floors and walls. Figure 5.2 shows the basic idea of a rising main system and Figure 5.3 shows a diagram of an internal fire barrier used to prevent the spread of fire to other floors.

When a switchboard or heavy piece of electrical apparatus arrives on site, it should be moved with care to a place where it cannot become damaged or soiled by a dusty/greasy working environment. It should also be protected against damp and corrosive conditions.

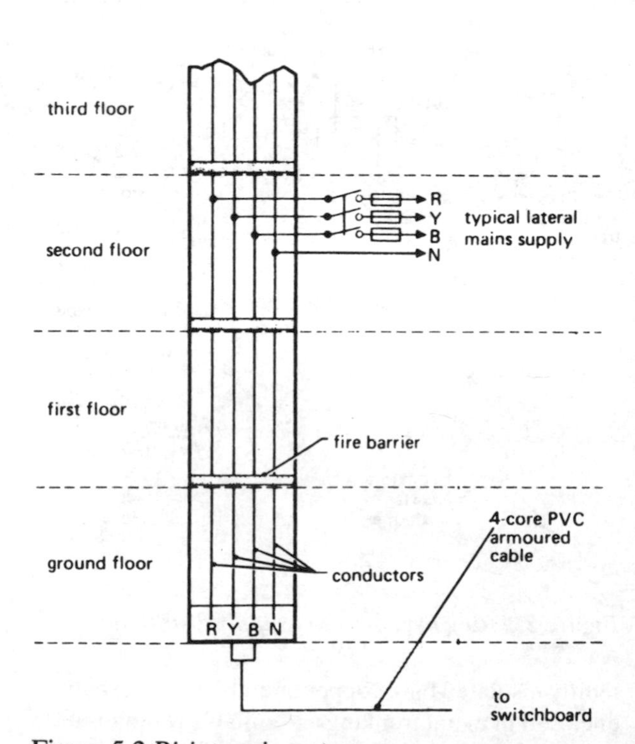

Figure 5.2 *Rising main system*

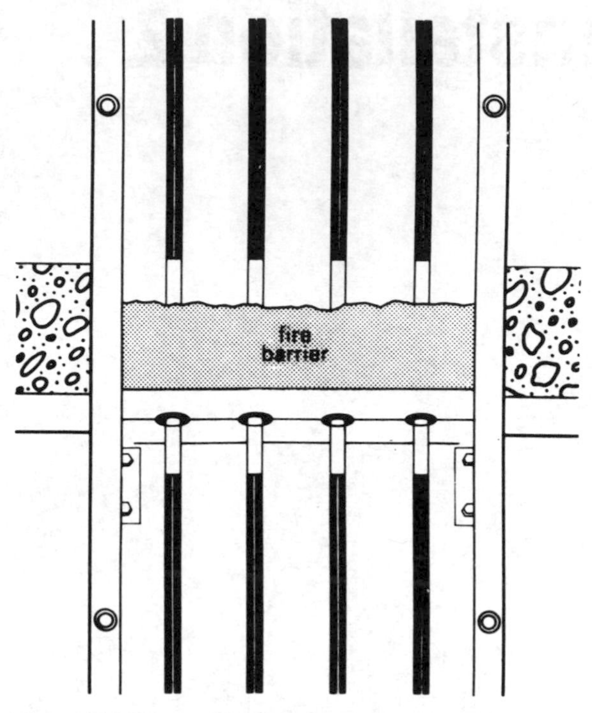

Figure 5.3 *Internal fire barrier*

It is worth mentioning the importance of sufficient information being given to manufacturers when ordering specially designed switchboards. Information must be given on the size of the incoming cable and the number of cores and whether or not the conductors are aluminium or copper. The manufacturer will want to know if access is at the back or front for ease of maintenance and what headroom is required to terminate the incoming cable; if this is from above or from below, and if the latter, whether there will be a trench available and what its dimensions are. Some industrial switchboards are made up on site and it is important to check with the manufacturers' latest catalogues when ordering components (see Figure 5.4).

Final circuits

Section 314 of the IEE Regulations is concered with *installation circuit arrangements*. There are four important regulations here which not only consider the reasons why the installation should be divided into final circuits having overcurrent protection, isolation

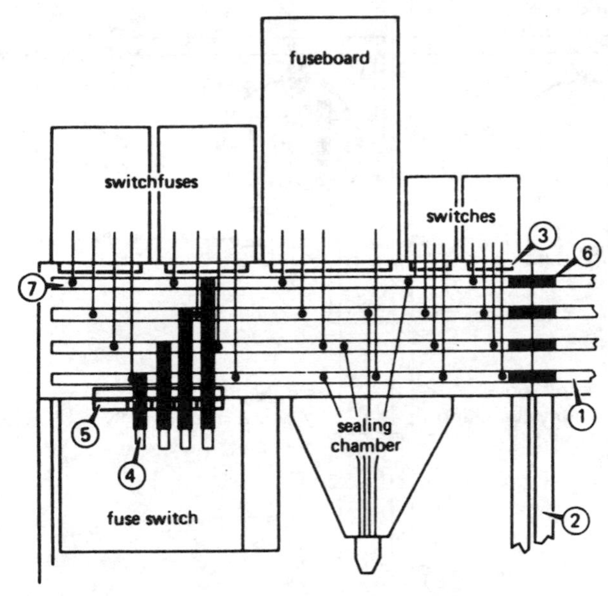

Figure 5.4 *Components of a switchboard*
1 Busbar chamber
2 Pedestal and backplate
3 Switch and fuseboard clamping channels
4 Fuse switch/busbar interconnections
5 Insulation plates
6 Busbar extension couplings
7 Busbar cable connections

and switching, but also considers the need for further separation to prevent indirect energisation.

A final circuit is basically an outgoing circuit fed from a distribution board intended to supply energy to current-using equipment. They are commonly divided into those for *lighting*, those for *heating* and those for *power*. This concept is, however, now becoming a little outdated, with more and more Category 2 circuits demanding separate consideration.

The correct procedure in selecting a cable for a final circuit was given in the section on installation design, once of course the wiring system was chosen, and this approach also considered the necessary protection against the two overcurrent conditions. The device used for overcurrent protection can also be used for *isolation* of the circuit, since a fuse link or circuit breaker is classified as an isolating device. The object of such isolation is to make 'dead' those parts of electrical equipment that are 'live' in normal use, so that work can be carried out. There are numerous IEE Regulation requirements to be found for isolation, as well as switching, and reference should be made to Reg. 13–06, Chapter 46 and Sections 476 and 537. In practice, there must be a means of isolation placed at the origin of every installation and it must be suitably placed ready for operation. This is particularly important when a supply is required in another building.

Where electrical equipment is fed from a final circuit, an isolator should be placed adjacent to the equipment, but where it is remotely placed, provision has to be made to stop it being inadvertently reclosed during operation. This is often achieved by providing it with a lock, key or removable handle.

It will be noticed from the IEE Regulations that some circuits might need additional forms of switching besides functional switching. Such switching is for *switching off for mechanical maintenance, and/or emergency switching*. The former relates to the protection of persons undertaking non-electrical tasks such as cleaning and replacement of lamps, whereas the latter relates to the rapid cutting off of a circuit or item of equipment from the supply in order to remove a hazard. It should be noted that a plug and socket cannot be used for emergency switching but that a single switch or combination of devices initiated by a single action (emergency push buttons) disconnecting the incoming supply can be used. Emergency switching devices are to be clearly identifiable and preferably red in colour. They must be manually operated and placed in a position of accessibility.

With reference to a domestic premises, the final circuits commonly found are: *lighting, heating, socket outlets and cooking appliances*. The normal practice for an upstairs/downstairs premises is to have two lighting circuits and two ring final circuits feeding BS 1363, 13 A socket outlets. The wiring system is likely to be PVC/PVC/c.p.c. cables to BS 6004. The lighting circuits will be wired by the ceiling rose loop-in method, with the cables passing through drilled wooden joists and terminating at switch positions from vertical drops. These downward drops from the ceiling will receive additional mechanical protection by being enclosed in PVC tubing or channelling.

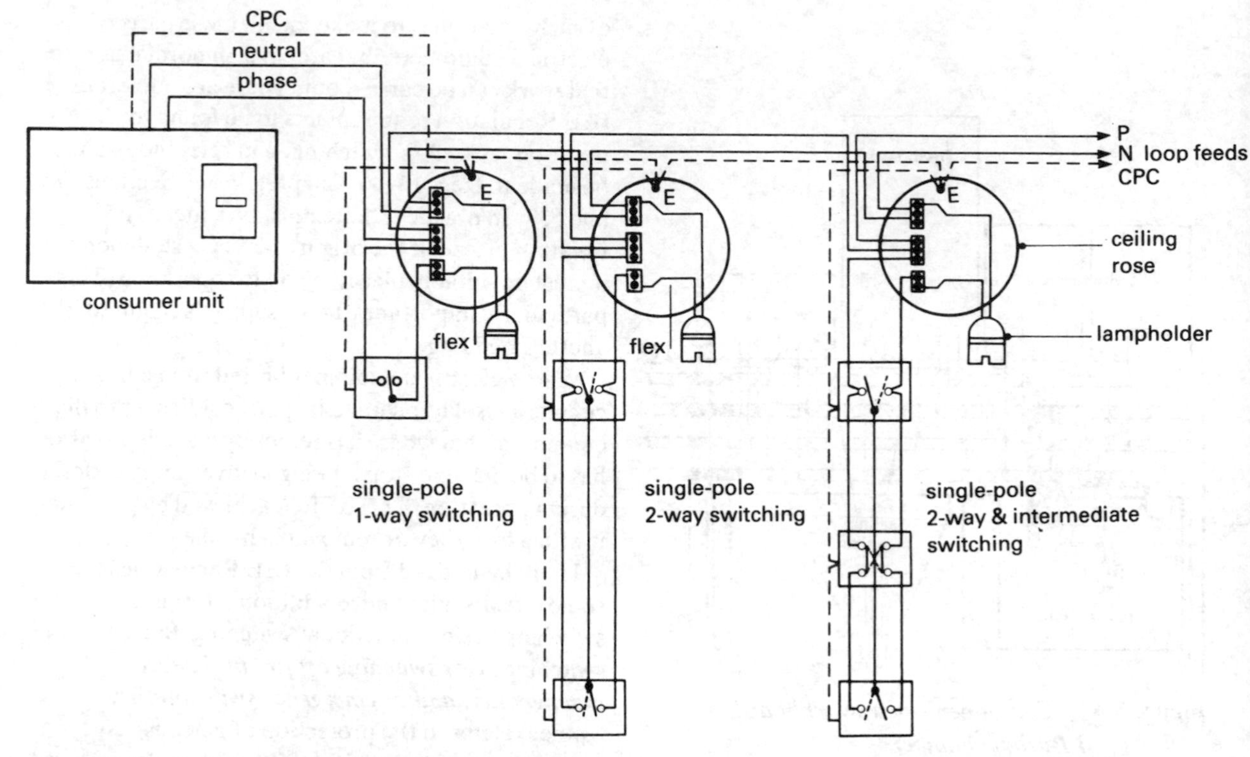

Figure 5.5 *Common methods of switching a lighting point*

There are numerous switching arrangements, some of which are shown in Figure 5.5. Other final circuits will be wired in a similar manner and it is important to keep up a standard of good workmanship, particularly with regard to fixings and terminations of the wiring system. The IEE Regulations should be consulted in this respect, especially Part 5. Some of the main considerations are listed below:

(a) Avoid mechanical stress on cables and conductors by providing adequate supports and allowing sufficient tails at termination points. Do not overtighten connections or cut strands out at terminal posts.
(b) Identify all conductors according to their colour coding, especially at outlet points where black conductors are often used for phase conductors and also where bare c.p.c. conductors are terminated.
(c) Use recommended materials made to British Standards or recognised foreign standard.
(d) Be aware of heat sources and sleeve conductors with appropriate heat resistant material. This is particularly important in some types of luminaire as well as space heaters, immersion heaters and electrical appliances such as cookers, irons, etc.
(e) Be aware of environmental conditions affecting the wiring system, such as dampness leading to corrosion, coldness leading to PVC becoming brittle, and hotness leading to distortion and possible deterioration of insulation.
(f) Take additional care with protective conductors, making sure that adequate continuity exists between metal boxes and fittings.
(g) Be aware of the expected dangers to persons in bathrooms and shower rooms and use pull-cord switches for controlling lighting and heating circuits. Use recommended lampholders and fittings and provide supplementary bonding to extraneous conductive parts.
(h) Create an order of wiring the phase, neutral and protective conductors inside the distribution board or consumer unit so that circuits can be easily traced. Leave details of circuit

destinations, as well as details of the person responsible for undertaking the work.

(i) Inspect, test and commission the electrical installation according to the requirements of the IEE Wiring Regulations.

(j) It should be good practice to provide a client with the necessary information on the wiring, such as plans showing cable routes and instructions on the operation of heating systems, alarm circuits and circuit protective devices, including residual current devices.

Final circuits in a factory or commercial premises will perhaps require a totally different arrangement from that described, although most of the points above are still applicable. One of the first considerations will of course be the design of the building and its use. Lighting might involve discharge lamps, and socket outlets may be designed to BS 196 or BS 4343 instead of BS 1363. The heating system may be combined with the ventilating system and there may be final circuits to motors, welders, lifts, cranes and other apparatus. The a.c. supply in most cases will be 415 V/240 V, but there may be equipment requiring a higher voltage than 415 V. At voltages above 240 V, there will be greater attention paid to warning notices and safety signs required by BS 5378. High-voltage discharge lighting will require a supplementary sign with the word *DANGER* indicated, as well as the highest open circuit voltage to earth (Reg. 554–02). Similarly, in a multigang lighting switch where more than one phase is present or in the case of switches or even socket outlets on different phases within arm's reach, a warning notice is required of the maximum voltage present (see Reg. 514–10–01).

Wiring systems

Chapter 52 of the IEE Wiring Regulations concerns selection and erection of wiring systems. In practice, there may be several factors likely to influence the choice of wiring system for a particular electrical installation; for example, the type of building and its use. A '*surface*' type wiring system may be acceptable in a factory or workplace but may be totally unacceptable in an office block or hotel or even domestic dwelling, simply on the grounds of appearance and aesthetic taste. Here, it is usual to install a hidden or '*flush*' type wiring system, such as PVC-insulated cables, mechanically protected under plaster. There is also consideration for the environment, such as excessive temperature, as well as any particular external influence from a corrosive atmosphere such as salt or dust. Some of these factors have already been mentioned. Questions will need to be asked about the chosen system's durability, its mechanical protection and cost comparison with other favourable systems; not only might there be a material cost benefit but also an installation time benefit. Another important question is whether the chosen system needs to cater for any likely alterations and/or the installing of additional circuits. Fortunately, in many large premises, one is likely to come across numerous wiring systems and these are often integrated with each other so that alterations and modifications can occur.

Some common types of wiring system found today are:

(a) PVC-insulated PVC-sheathed cables,
(b) PVC-insulated armoured PVC-sheathed cables,
(c) mineral-insulated metal-sheathed cables,
(d) metal and plastic conduit systems (incorporating cables),
(e) metal and plastic trunking systems (incorporating cables),
(f) busbar trunking systems.

It should be pointed out that there are various support arrangements for the above systems, such as cable tray, ducting and cable trench; see the methods in Table 9A of the IEE Wiring Regulations.

PVC-insulated PVC-sheathed cables

This wiring system shown in Figure 5.6 has a general use and will be found listed in Table 4D1–2 of the IEE Regulations. The common PVC/PVC/c.p.c. is widely used for surface wiring, where it needs to be clipped and supported to meet the requirements of the Regulations. There are requirements for the internal radii of bends and supports. The wiring

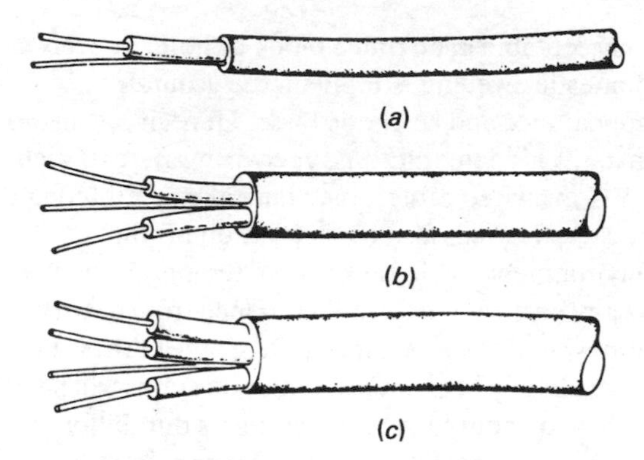

Figure 5.6 *PVC insulated, PVC sheathed cables 600 V/1000 V*
(a) Single core
(b) Two core
(c) Three core

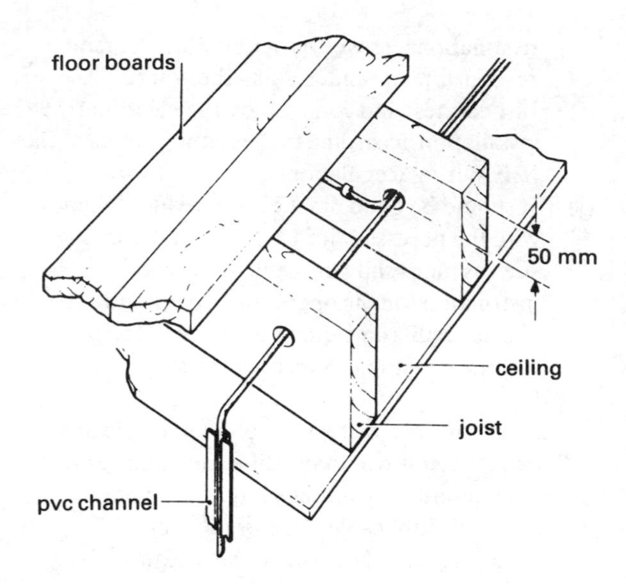

Figure 5.7 *PVC cables under floorboards*

system can be hidden from view, above a false ceiling, in joists or buried under plaster. In a false ceiling, the cables should be kept clear of sharp edges and be adequately supported. Where they are run through holes in wooden joists, the cables must be at least 50 mm from the top or bottom of the joist (Reg. 522–06–05). If cables are to be laid in existing notches, they should be provided with mechanical protection to prevent damage occurring from floor fixings. Figure 5.7 shows a typical arrangement through joists. It should be noted from Table 4D2 that these cables have a maximum conductor operating temperature of 70°C, and it will be observed that various correction factors apply for ambient temperatures in excess of 30°C. Chapter 42 of the IEE Regulations deals with *protection against thermal effects*. It is generally recognised that a group correction factor does not apply to domestic final circuits but, where this is the case in other instances, Table 4B1 has to be used. At terminations, the outer PVC sheath should not be removed any further than is necessary and all conductors should be identified by their appropriate colour coding. Sections 526 and 527 of the IEE Regulations cover *accessibility* and *selection* and *erection* to minimise the spread of fire respectively. It is important to make sure that no mechanical stress is placed on the cables and that cable glands securely retain the outer sheath of the cable. On no account should conductor strands be cut in order to fit into a termination post.

PVC-insulated armoured PVC-sheathed cables

PVC armoured cables have a wide commercial and industrial application and there are several tables in the IEE Regulations concerning both the copper and aluminium conductor types. Figure 5.8(a) shows a diagram of a typical cable; the armouring consists of galvanised steel wire secured between PVC bedding and a tough PVC outer sheath. The armouring is often used as a protective conductor but this should be checked by calculation using Reg. 543–01–03 as previously discussed. An alternative method is to apply the following table, which is derived from BS 6346 and shows the nearest smaller copper cable size equivalent to the armouring of the chosen cables.

Nominal area of conductor (mm^2)	*Nominal area of copper conductor equivalent to armouring* (mm^2)		
	2-core	3-core	4-core
2.5	1.5	1.5	1.5
4.0	1.5	2.5	4.0
6.0	2.5	4.0	4.0
10.0	4.0	4.0	4.0
16.0	4.0	4.0	6.0
25.0	6.0	6.0	6.0
50.0	6.0	6.0	10.0

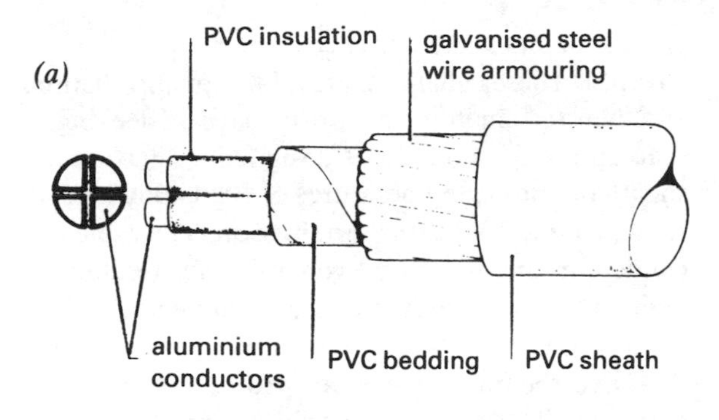

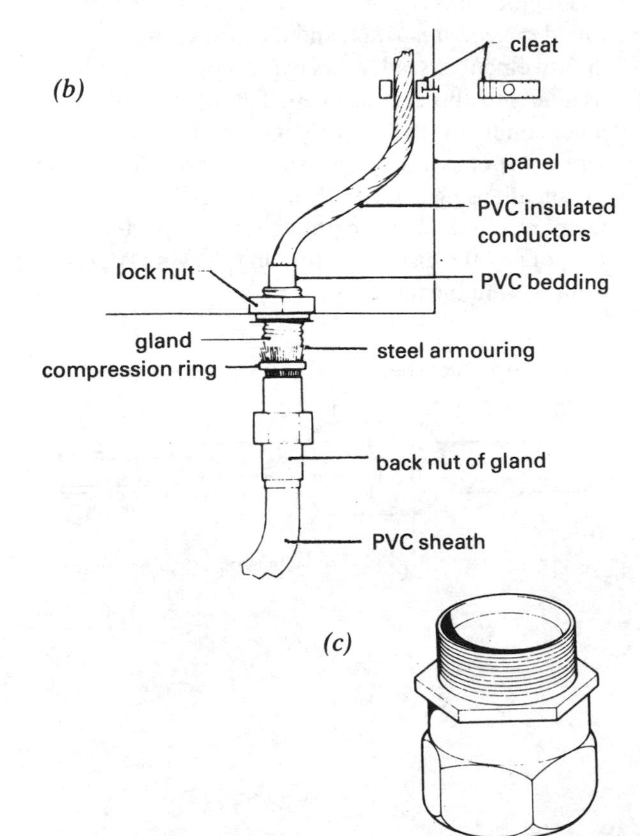

Figure 5.8(a) *PVC armoured PVC insulated cable; (b) termination; (c) gland*

The cables are often clipped on a surface using recommended wall cleats, or they can be installed on cable tray or run in a trench as outlined in Table 4A of the Regulations. Where they are buried in soil, one will find that the usual practice is to provide either cable tiles or yellow warning tape laid on top so as to denote their position. They must also be installed deep enough to avoid damage from any possible ground disturbance (see Reg. 522–06–03).

It is important to see that cable terminations are mechanically and electrically sound and that the cable gland and steel armouring make an effective earth connection. The use of an earth tag is recommended in order to give earth continuity between the gland and the steel enclosure to which the gland is fixed. All conductor cores of multicore cables should be identified properly using appropriate markers.

It was mentioned under the section on earthing arrangements that a PEN conductor was used in a TN–C–S system. There are several types of armoured cable chosen for this system, designed on a *combined neutral earth* (c.n.e.) basis. Briefly, they come under the names of *consac*, *districable*, *waveformal and unscreened*, and they are illustrated in Figure 5.9.

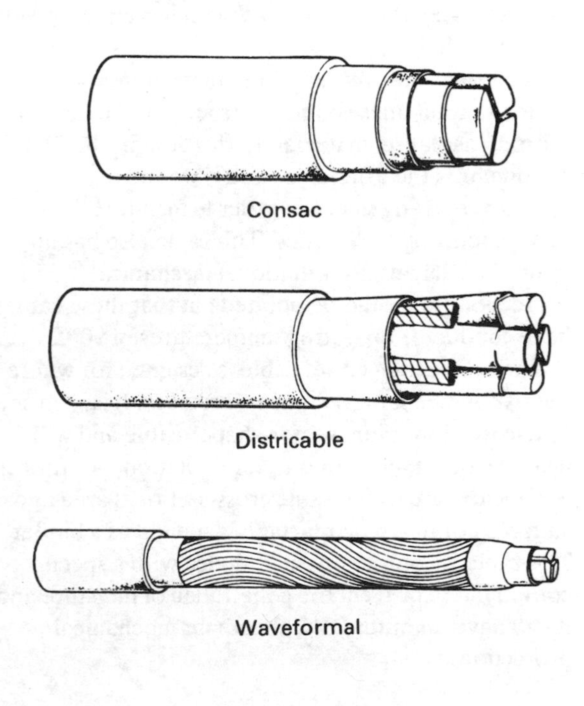

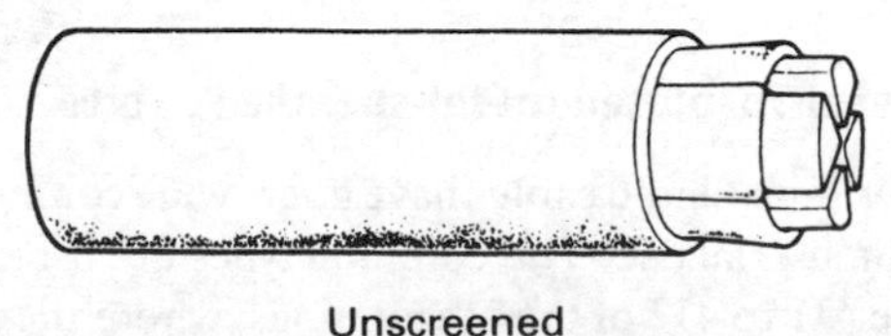

Figure 5.9 *Four types of c.n.e. cable*

It will be seen that the three-phase conductors of *the consac* cable are sector-shaped and made of aluminium. They are insulated with impregnated paper tapes and identified with the numbers 1, 2 and 3. The extruded aluminium sheath is designed to give adequate conductivity for use as a PEN conductor, and this is protected by a layer of bitumen and a further PVC outer sheath.

The *districable* again has sector-shaped aluminium phase conductors, which are insulated with a high-temperature cross-linked polyethylene (x.l.p.e.) material. Its PEN conductor may, however, be circular and covered with lead to give it excellent corrosion-resistant properties when buried in the ground. The conductor cores are laid up together with polypropylene wormings and protected by two steel tapes in intimate contact with the PEN conductor throughout the length of the cable. This provides a *low impedance path* for any fault current that may arise and, like the previous cable, it has an extruded PVC sheath.

In the *waveformal* cable, the three aluminium sector-shaped phase conductors are surrounded by x.l.p.e. insulating material. In this design, the PEN conductor is the armouring and is wound in a *sinusoidal* arrangement in order to facilitate 'live' service jointing techniques. This cable also has an outer PVC sheath for additional mechanical protection. It should be pointed out that these cables have conductor operating temperatures of 90°C.

The *unscreened* c.n.e. cable is designed for where the risk of mechanical damage is remote. It is seen as a four-core aluminium sector-shaped cable and will again be provided with x.l.p.e. insulation. All four conductors are of the same cross-sectional area and a suitable non-hygroscopic tape is applied as a binder. The cable is completely surrounded with a special compound to prevent the penetration of moisture and it will have an outer PVC sheath for mechanical protection.

Mineral-insulated metal-sheathed cables

Mineral-insulated cables have a very wide commercial and industrial use. The common types are listed in Table 4J1 to 4J2 of the Regulations, where they are divided between light-duty use and heavy-duty use and the outer sheath either covered with PVC or not covered. These cables are ideal for 'hot' installations and their termination accessories such as sleeving, seals and compound can be designed for maximum sheath operating temperatures of 105°C. Figure 5.10 shows a typical MI cable termination. The cable conductors and sheath are generally constructed of high-conductivity copper or aluminium and, as already indicated, both types may be provided with a PVC oversheath to give added protection against corrosion. The insulation medium between the conductor cores is a compressed mineral powder called *magnesium oxide*, and there have been improvements to reduce its hygroscopic nature (i.e. its ability to absorb moisture). Termination of the inner conductor cores involves several tools. The screw-on pot seal is widely used and this is attached to the sheath using a pot wrench. The correct temperature sealing compound is then inserted into the pot and the disc and insulating sleeving attached using a crimping tool.

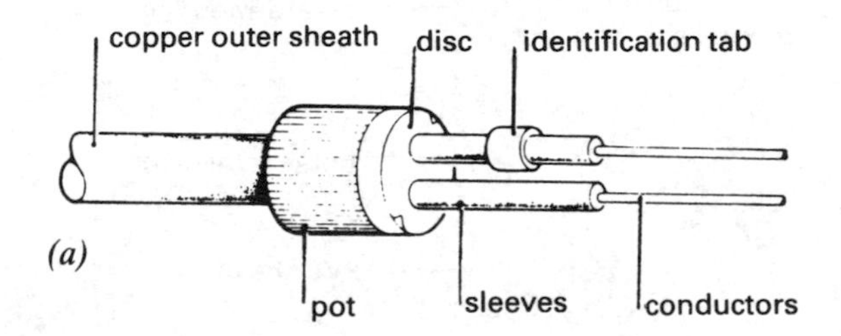

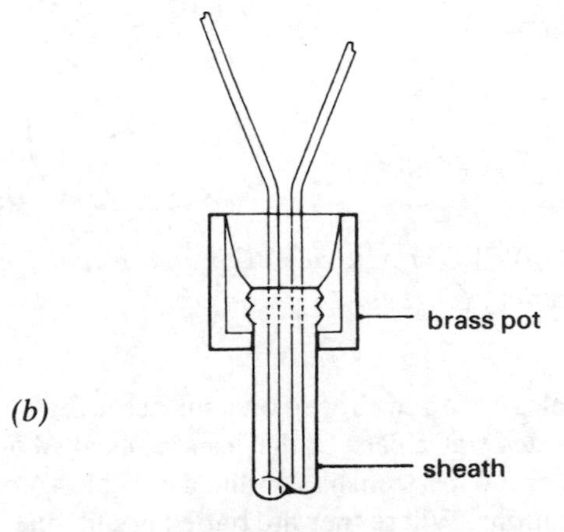

Figure 5.10 *(a) cable termination*
(b) screw on seal

The cables are often referred to as mineral-insulated metal-sheathed cables (m.i.m.s.) and are fireproof, waterproof and oilproof. They are also non-ageing and have higher current ratings compared with equivalent cable sizes. They are ideal for fire-alarm circuits, boiler rooms and garage/petrol filling installations where dangerous atmospheres are likely to be present in Zone 1 hazardous areas. Here, they are fitted with flameproof glands and the cables have an overall extruded covering of PVC. The metal sheath of these cables satisfies both shock protection and thermal protection requirements since it is approximately four times the cross-sectional area of the inside cores. To satisfy earthing connections, earth-tail pots are available having a cross-sectional area equal to the related size of the phase conductor(s). It is recommended that these should always be used for continuing the protective conductor function of the sheath through the earthing terminal at outlet points.

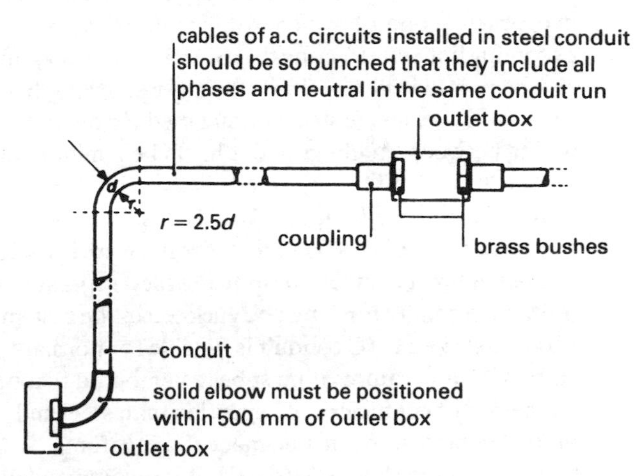

(a) Conduit requirements

Metal and plastic conduit systems

Conduit systems are in wide use today either as surface or flush wiring arrangements. The common sizes are 16 mm, 20 mm, 25 mm and 32 mm. They provide the installed cables, such as the single-core PVC-insulated or ethylene propylene rubber (e.p.r.)-insulated cables, with additional mechanical protection. The advantage of conduit is the amount of flexibility one can have with final circuits, since cables can be added or withdrawn. It is important to see that the conduits are erected first before the cables are drawn in (Reg. 522–08) and it is equally or more important that *metal conduit* systems are both electrically and mechanically sound, even if it is decided to install protective conductors for additional safety (Reg. 543–02). Further requirements are made in Chapter 54 (see Figure 5.11).

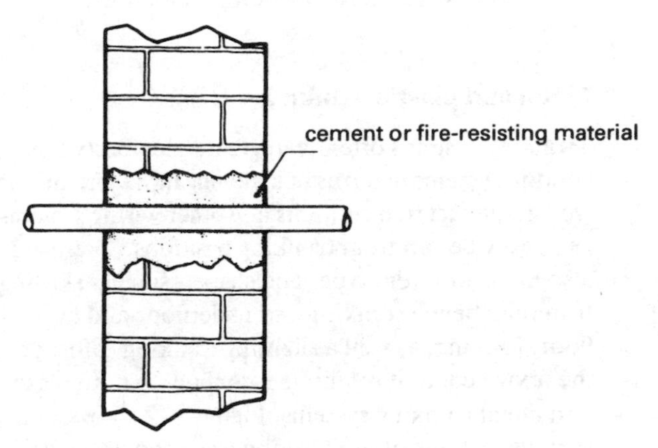

(b) Conduit passing through a wall

Conduits should be distinguished from other services by the colour orange (Reg. 514–03–01); in damp situations, galvanised conduit or plastic conduit should be used. With either method, it is essential to provide drainage facilities for the release of moisture (see Reg. 522–03–02). One of the most important requirements is Reg. 521–03 (*conductors of a.c.*

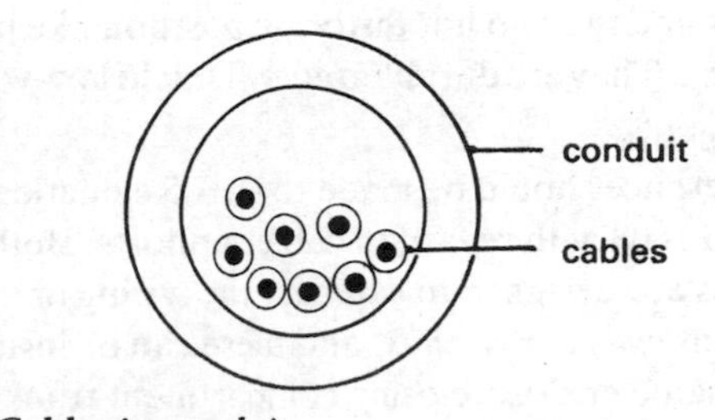

(c) Cables in conduit

Figure 5.11

circuits installed in ferrous enclosures); the reason behind this regulation is to stop *eddy currents* flowing in the metal conduit, which would result in them becoming hot and affecting the cables inside. It should also be pointed out that 3 m is the maximum length of span for heavy guage steel conduit used for overhead wiring between buildings and it has to be a minimum size of 20 mm and be unjointed.

With regard to *plastic conduit*, the system is of course non-rusting and is a protective measure in itself against indirect contact. Despite this fact, separate protective conductors must be enclosed in the system and whilst rigid PVC conduit is suitable for normal ambient temperatures it must be remembered that it has a greater coefficient of expansion than steel and therefore provision must be made to allow for movement (see Reg. 522–01–02). It is recommended that expansion couplers be fitted in the worst affected areas. Reg. 522–05 should be noted with regard to plastic boxes used for suspending luminaires.

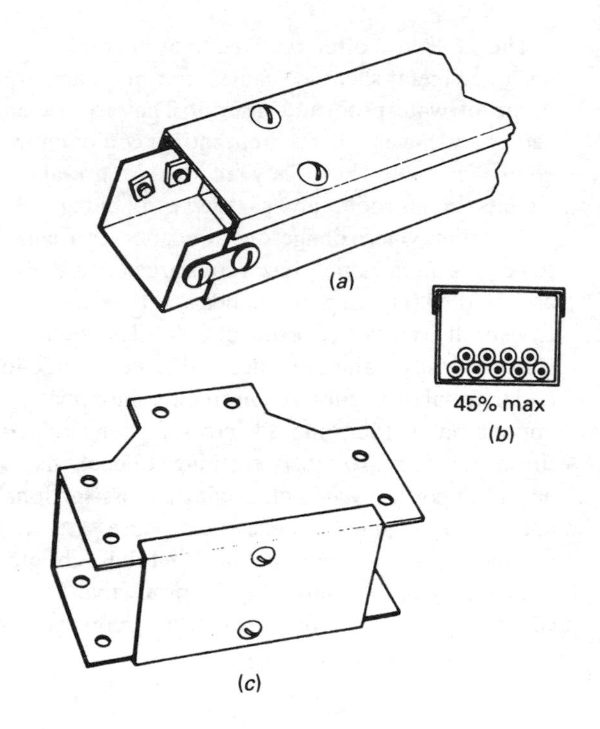

Metal and plastic trunking systems

Trunking systems offer even greater flexibility than conduit systems in terms of installing more circuits, as well as the fact that conduits and other wiring systems can easily be run from trunking positions. There are also many different types and shapes, such as skirting trunking, bench trunking and underfloor and flush floor trunking, as well as lighting trunking. Most of the above conduit requirements apply to both plastic and metal trunking systems. Figure 5.12 shows various sections of steel trunking and the use of a copper link for added earth continuity, but it is worth noting that steel trunking systems actually provide an effective protective conductor for numerous final circuits and they do not distort in position like plastic trunking. They are durable and will shield low-voltage data signals.

Reference should be made to IEE Regulations, Reg. 522–08 with regard to cable supports. Both systems are designed to cater for the wiring of different *category circuits*, and these can be installed in the same enclosure using compartment trunking. The requirements for category circuits can be found in Reg. 528–02. Probably one of the most important factors in choosing plastic trunking is the protection it

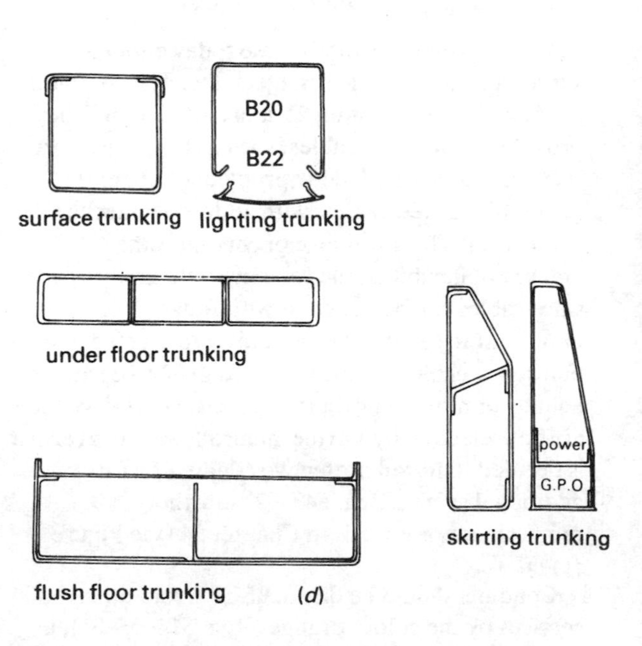

Figure 5.12 *(a) Standard trunking showing copper earth link*
(b) Cables in trunking
(c) Tee unit
(d) Common types

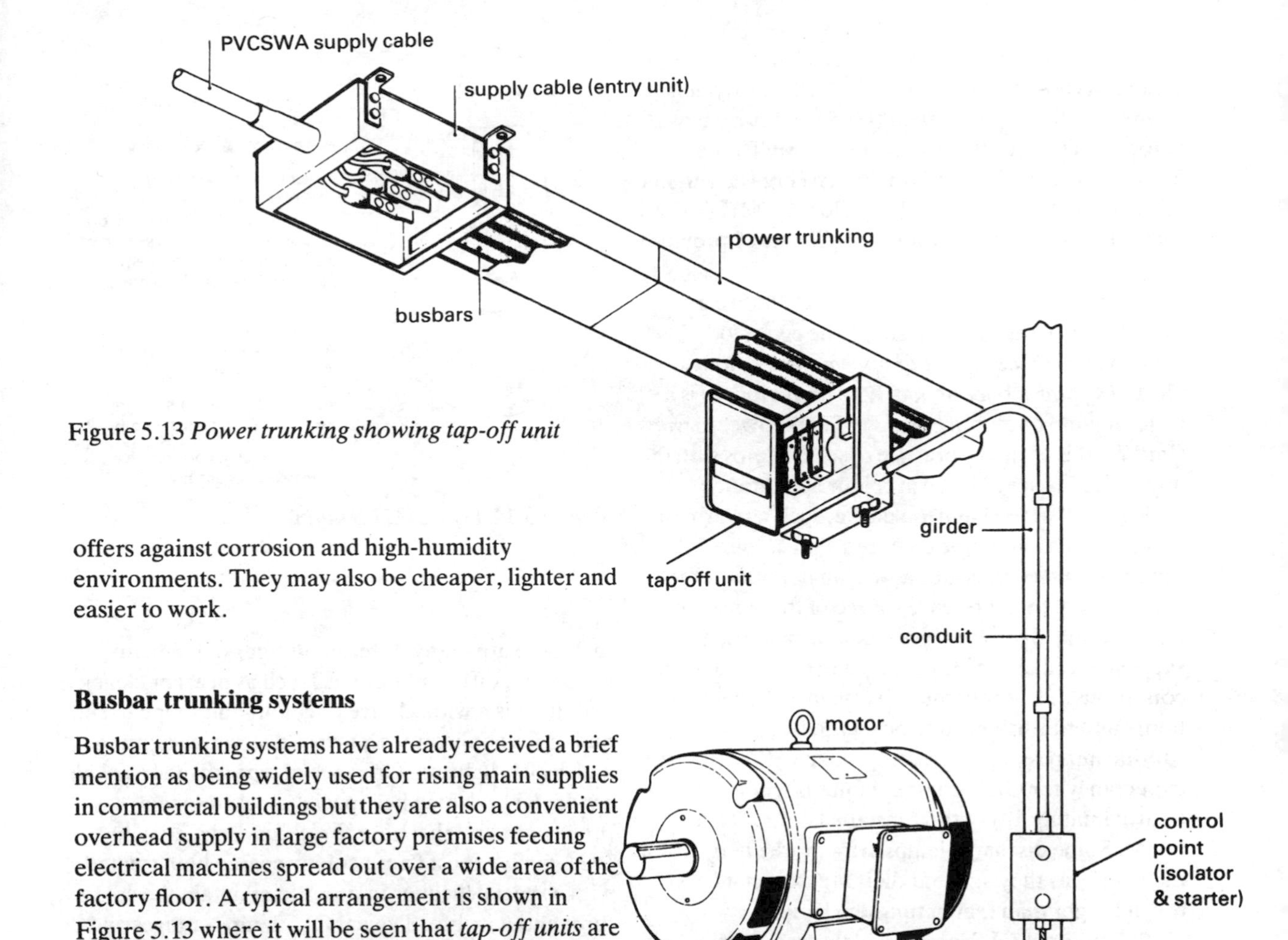

Figure 5.13 *Power trunking showing tap-off unit*

offers against corrosion and high-humidity environments. They may also be cheaper, lighter and easier to work.

Busbar trunking systems

Busbar trunking systems have already received a brief mention as being widely used for rising main supplies in commercial buildings but they are also a convenient overhead supply in large factory premises feeding electrical machines spread out over a wide area of the factory floor. A typical arrangement is shown in Figure 5.13 where it will be seen that *tap-off units* are fitted for the final circuit wiring. These units are generally spaced between 0.5 and 1 m apart and they are normally fused at ratings of 30, 60 and 100 A. The units are plugged in by hand and firm contact is made with the bare busbars by physically tightening up wing nuts. On some tap-off units, special spring contacts are provided to ensure that the casings are earthed before contact is made to the 'live' bars. The internal conductors are generally made of high-conductivity copper bars of rectangular section and these are supported at regular intervals by bakelite insulators. The main supply cable entry points are often situated at the ends or in the centre of the trunking length itself and these lengths are supported by brackets fixed to the trusses of the building. As mentioned before with ordinary metal trunking, additional earth continuity may be provided by fitting copper bonding strips between each length.

When ordering trunking busbar systems, manufacturers often want to know installation details, such as: current rating, type of cable feed, thickness of floor slabs, number of interfloor barriers, etc. Expansion joints and other accessories are normally supplied as part of the complete installation.

Lighting installations

Several switching arrangements have already been given for lighting circuits but one needs to know a little more about the different types of lamp, their requirements and characteristics to appreciate the reason for their choice in an electrical installation. In general, types of lamp can be classified as: (a) *filament lamps*, i.e. tungsten filament (GLS) and tungsten-halogen (TH); (b) *fluorescent lamps* (MCF), i.e. tubular and compact; and (c) *gas discharge lamps*, i.e. mercury (MBF/U), (MBFR/U), (MBTL/MBTF) and sodium (SOX), (SON), (SON-T). Chapter 1 provides lamp data.

When a type of lamp is selected for a lighting installation, consideration should be given to: *efficacy, light output, colour appearance, colour rendering and service period*. The term efficacy is a ratio of lamp output in *lumens* to lamp power in *watts* (lm/W). It is often the practice to select lamps with the highest efficacies. The term colour appearance indicates what the lamp looks like, such as warm or cool, and it must not be confused with colour rendering which indicates how a lamp renders objects compared with a reference source of the same colour temperature. However, good colour rendering often suggests similarity of appearance to that of daylight conditions. The economic life of lamps is stated in hours and this will need to be considered in any routine lamp replacement programme. The usual life expectancy for GLS tungsten lamps is when approximately 5% of their installed number has failed. Some discharge lamps draw more energy from the supply as they age, but their life expectancy is much longer than that of tungsten lamps.

The tungsten GLS lamp is widely used, particularly in domestic lighting circuits, and a diagram of its construction is shown in Figure 5.14.

The lamps have watt ratings ranging from 15 to 1000 W and they may have different types of cap such as Bayonet Cap (BC), Edison Screw (ES) or Goliath Edison Screw (GES); GES is used for 300, 500 and 1000 W lamps. Most GLS lamps are gas filled to reduce the rate of filament evaporation, which allows them to operate at much higher temperatures (~3000°C). Lamps can be made with a single-coil filament or a coiled-coil filament. The latter are preferred since they have a higher light output. The lamps can be operated in any position but they last longer in the vertical cap-up position. Life expectancy can be as high as 2500 h for double-life-plus lamps but, while colour rendering is good and their low cost an advantage, they suffer from having a poor efficacy compared with other types of lamp. It should be noted

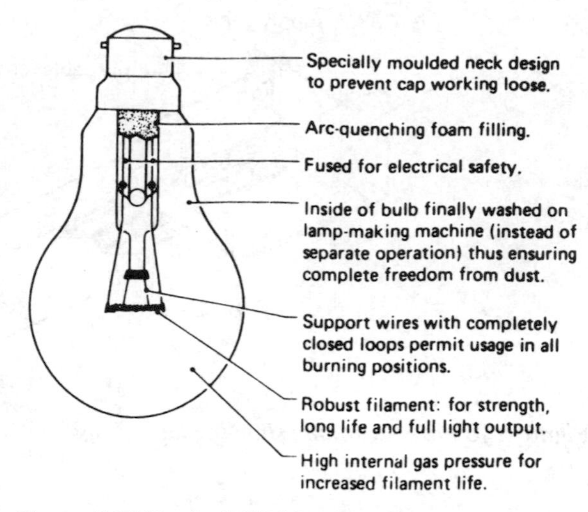

Figure 5.14 *Typical GLS lamp*

that there are many different designs of filament lamps with different finishes (such as pearl or clear), and there is a wide choice of colour and shape to suit numerous applications, such as decorative lighting and display lighting. One special type is called a *rough service lamp*, which has special shock-resistant properties; and there are *firelamps* designed with a BC 3-pin having a flame-colour-coated bulb treated with a special lacquer in order to maintain the colour throughout its life. Typical GLS lamp wattages and lighting design lumens (LDL) are given in the table below:

Single-coil lamp (240 V)		*Coiled-coil lamp (240 V)*	
Watts	*LDL*	*Watts*	*LDL*
15	150	40	390
25	200	60	665
40	325	75	885
60	575	100	1260
100	1160	150	2075
150	1960		
200	2720		
300	4300		
500	7700		
750	12 400		
1000	17 300		

Calculation of efficacy for a 100 W lamp:

$$\text{Efficacy} = \text{lumens/watt} = 1160/100 = 11.6 \text{ lm/W}$$

Tungsten-halogen lamps

These lamps were introduced in the 1950s. Two particular designs are shown in Figure 5.15. For their operation the tungsten filament is enclosed in a gas-filled quartz tube together with a carefully controlled amount of halogen, such as iodine or bromine. When the tungsten filament is heated by an electric current, the halogen vaporises and controls the evaporation of the filament. Tungsten vapour is carried to the comparatively cool wall of the bulb by convection currents, where it combines with the halogen to form a tungsten halide. This compound then returns to the filament, where the very high temperature chemically converts it back into tungsten and halogen.

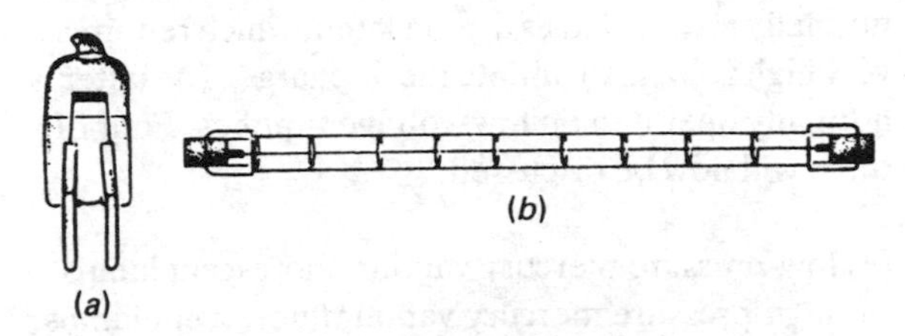

Figure 5.15 *(a) Single-ended tungsten-halogen lamp (b) Linear double-ended tungsten-halogen lamp*

The cycle is continually repeated and performs a self-cleaning action on the inner surface of the bulb. In a standard GLS filament lamp the evaporation of the tungsten causes a blackening of the bulb after a period of use, but in the tungsten-halogen lamp this is completely eliminated. The lamp has a much higher efficacy and longer life than the standard GLS lamp. However, although it is a compact and easily controllable light source, it should be pointed out that a minimum bulb-wall temperature of 250°C is needed to maintain the tungsten-halogen cycle. Also, to prevent oxidation of the lead-in wires and failure of the pinch seal, the temperature must not exceed 350°C.

The two types of lamp shown are called *linear double-ended lamps* and *single-ended lamps* (M class). The linear lamps have a higher rating and must be operated in the horizontal position (or within 4° of the horizontal). If the angle is too steep, halogen vapour will migrate to the lower end, leaving the upper end starved. This will result in rapid blackening of the bulb and, correspondingly, reduced life.

Linear lamps are widely used for display lighting, particularly floodlighting (Figure 5.16) and office copying equipment requiring a linear light source. Single-ended lamps have various ratings up to 500 W and again can be used for display lighting, studio and theatre lighting, spotlights and traffic signals.

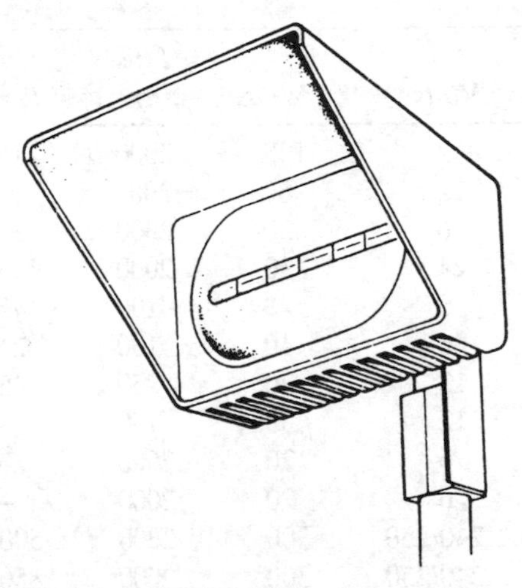

Figure 5.16 *Sun 500 floodlight showing linear tungsten-halogen lamp*

Details of the linear and Class M type of lamps are given below. It should, however, be pointed out that halogen lamps must be handled carefully, especially when being fitted. It is important not to contaminate the outside surface of the quartz tube with dirty or greasy hands, as this will result in fine cracks appearing on the bulb, causing premature failure. In practice, it is advisable to use a paper sleeve over the lamp, or handle the lamp by its ends. If it is accidently handled, it should be cleaned with a solvent such as industrial spirit, carbon tetrachloride or trichlorethylene.

Linear double-ended lamps

Type	Volts	Watts	Life (h)	No. of lumens	Burning position
K9	240/250	300	2000	5000	horizontal
K1	240/250	500	2000	9500	horizontal
K2	240/250	750	2000	15 000	horizontal
K3	240/250	750	2000	15 000	horizontal
K4	240/250	1000	2000	21 000	horizontal
K5	240/250	1500	2000	33 000	horizontal
K8	240/250	2000	2000	44 000	horizontal
K6	240/250	2000	2000	44 000	horizontal

Class M single-ended lamps

Type	Volts	Watts	Life (h)	No. of lumens
M28	12	100	2000	2150*
M32	12	50	2000	900
M34	6	20	2000	350
M36	24	250	2000	5750*
M30	6	20	100	420
M29	6	10	100	210
M35	12	20	250	450
M37	12	55	750	–
M39	6	20	2000	–
M41	6	20	2000	–
M38	240/250	300	2000	5000
M40	240/250	300	2000	8500

Notes:
* The burning position of these two lamps is base down to base horizontal. All other lamps can be operated in any position.

The M35 can be operated at 13.2 V and the M38 and M40 are designed for other working voltages. The efficacy of the M40 is 28.3 lm/W compared with the linear K9 which is 16.67 lm/W.

Discharge lamps

Unlike incandescent lamps, discharge lamps have no filaments. Their light is the result of excitation of a gas or vapour inside a sealed glass tube containing two electrodes. When a typical lamp is connected in circuit with the supply, the voltage applied across its electrodes will cause ionisation of the gas or vapour filling. Ionisation is the excitation of gas atoms such that they become converted into ions.

Positive ions move towards the *cathode* (the negatively charged electrode) and negative ions move towards the *anode* (the positively charged electrode). On a.c., this conversion occurs every half cycle. The applied voltage causes ions to move at speed along the discharge tube and they either collide with neutral atoms, with each other or pass through the tube itself. They release their acquired energy, either in the form of heat or electromagnetic ratiation—this may be in the ultraviolet or infrared regions.

Continued ionisation must, however, be checked, otherwise unstable conditions may occur and give rise to increased lamp current. In practice, the problem is overcome by inserting an inductor or high-leakage transformer in the lamp circuit.

Generally speaking, there are two kinds of discharge lamp; one is called the *cold-cathode* type and the other the *hot-cathode* type. The former is normally referred to as a neon lamp, which requires a very high voltage to initiate the discharge. The latter is normally operated on low-voltage supplies. Four of these will now be discussed:

(a) low-pressure mercury vapour fluorescent lamps,
(b) high-pressure mercury vapour fluorescent lamps,
(c) low-pressure sodium vapour lamps,
(d) high-pressure sodium vapour lamps.

Low-pressure mercury fluorescent lamps

These are commonly known as *tubular fluorescent lamps* and they operate at low pressure. They are coded MCF, and Figures 5.17 and 5.18 show the constructional features of a typical lamp. Basically, it is a clear glass tube filled with a rare gas such as argon or krypton and a measured amount of mercury vapour. The inside of the tube is coated with a fluorescent phosphor, which absorbs ultraviolet radiation and transforms it into visible light. Different phosphors are used to emit light of almost any desired colour. At both ends of the tube are cathodes (Figure 5.18), formed out of a single or multiple coil of tungsten wire. These cathodes are coated with a special electron-emitting material, mounted on glass pinches and sealed into the ends of the lamp.

Fluorescent tubes are made in various lengths ranging from 150 to 2400 mm, with respective

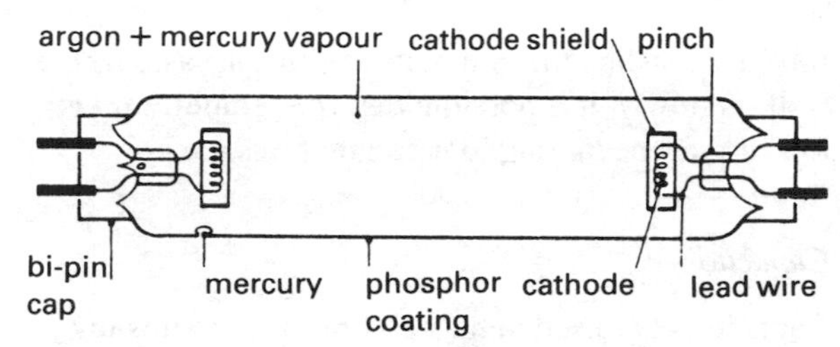

Figure 5.17 *Basic construction of a fluorescent tube*

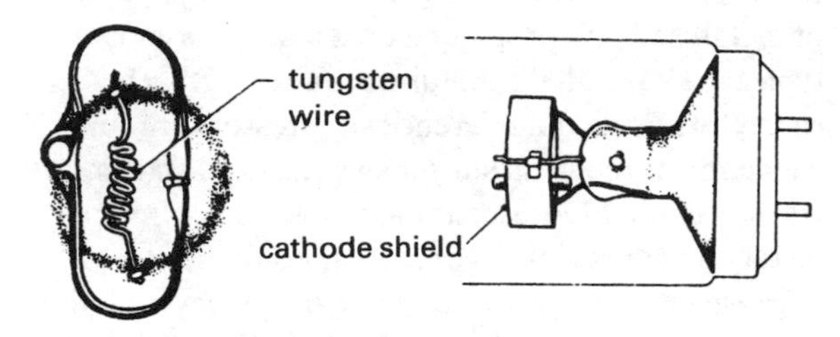

Figure 5.18 *Cathode of a fluorescent tube*

wattages ranging from 4 to 125 W. The majority of tubes are fitted with bi-pin lamp caps, with miniature bi-pin types for smaller tubes. For general lighting at maximum efficiency, the colour *white* provides a fair colour rendering with intermediate colour appearance. It has a lighting design lumens of 8800 lm and an efficacy of 70 lm/W.

Before a fluorescent tube will strike, its gas filling must be ionised. Normally, the voltage required to do this needs to be slightly higher than the voltage required to maintain the actual discharge. Since most lamps usually operate on a.c., the method of providing the initial voltage *surge* is by an inductive coil called a *choke*. This serves a dual purpose as it also limits the current through the lamp—the lamp's resistance decreases as the discharge increases. Unfortunately, the choke causes a lagging power factor, so a power factor correcting capacitor is incorporated in the control gear. Two common methods of starting a fluorescent lamp are given below.

Switch start circuit

Figure 5.19 is a typical switch start circuit using a *glow-type starter switch* (Figure 5.20). The starter switch is wired across one side of the lamp and its contacts (two bimetallic strip electrodes) are normally open.

When the supply is switched on, full mains voltage is applied across the starter contacts and, because of their closeness, a glow discharge takes place between them. The contacts quickly heat up and touch together, allowing current to pass through the lamp cathodes. The lamp will be seen lighting at either end before the arc actually strikes. The arc will strike only when the starter contacts cool down and part, breaking the choke's inductive circuit to create the necessary voltage surge. The voltage across the lamp is not sufficient to restart the glow discharge in the starter unit, so its contacts remain open.

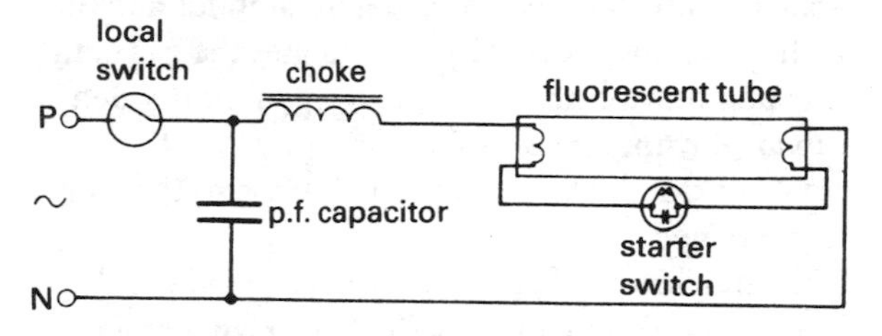

Figure 5.19 *Glow-type starter switch circuit*

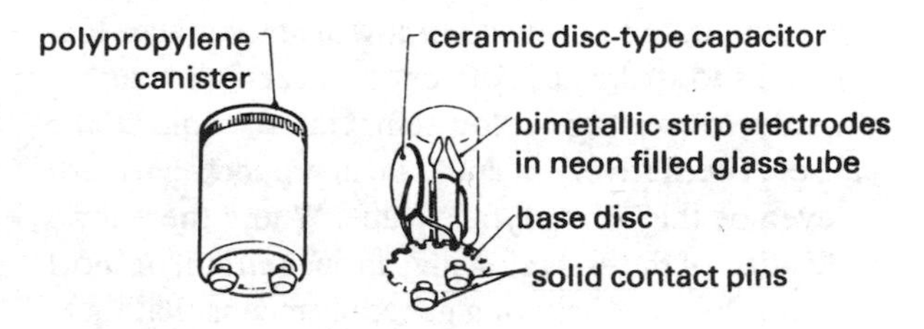

Figure 5.20 *Glow-type starter switch*

If the tube does not strike first time, then the process is repeated. Sometimes a faulty starter switch may open before the lamp cathodes have had a chance to heat up. This *cold* striking of the arc shortens the lamp life by eroding the electron-emitting material around the cathodes. Any reduced emission will lead to the lamp flashing *on* and *off*.

Complete tube failure occurs when the material ceases to produce electrons in sufficient quantity. This is often seen by severe blackening at both ends of the tube. Occasionally, the contacts of a starter switch may fail to open, resulting in the ends of the lamp glowing with no attempt to strike. It is important to use the correct starter switch for the size of tube. The simplest test of a starter switch is to insert it into a sound fitting, but a test rig comprising a starter-switch holder wired in series with a 25 W, 240 V tungsten filament lamp can be used. If the lamp flashes *on* and *off*, then the starter is working properly.*

Transformer quickstart circuit

This circuit is shown in Figure 5.21. It will be seen that the transformer is connected across the lamp cathodes. The discharge occurs as soon as the electrodes reach their correct operating temperature. When the pre-heating period is complete, the capacitive effect between the lamp cathodes and the earthed metalwork of the fitting ionises the gas in the lamp and the arc strikes. Some lamps are provided with an earthing strip and it is important for the metalwork of the fitting, especially close to the lamp, to be earthed.

The quickstart circuit has several advantages over the switch start circuit, namely that it provides an immediate start, no problem with faulty starter units, and no danger of cold starting. The transformer is a little more expensive and some difficulty may occur at low ambient temperatures or low mains voltage. If a tube fails to strike but both ends are seen glowing brightly, low voltage or low temperature could be the cause. Alternatively, it could be improperly earthed or even be the wrong type of tube. Where there is no indication of the ends glowing, an open circuit or short circuit is likely. Flickering generally means that the tube has reached the end of its life but, in cases where it starts slowly, it is possible that the temperature is low or that the lampholder connections are poor.

Capacitors

Capacitors are used in fluorescent tube circuits for both power factor correction and radio interference suppression. Power factor capacitors are normally of the dry-film construction, made by winding two strips of metallised polypropylene on a plastic former. A typical 125 W, 2400 mm tube requires a 7.2 μF capacitor. Radio interference suppressors are small capacitors fitted in all starter switches and starter transformers. Fluorescent tubes should not be installed near radio sets or similar equipment, because the tubes produce a high-frequency radiation signal that may be picked up by the radio aerial, especially in the medium-wave band. The signal normally occurs in a particular tube's cathode support wires, or close to it, but it does move up and down the cathode owing to loss of emissive material.

Methods of overcoming the problem of radio interference have been to install the aerials outdoors, reverse the fluorescent tube, or provide the circuit with a radio filter network. Typical capacitor values here are about 0.005μF.

Stroboscopic effect

In installations where rotating machinery is present and where discharge lamps are used, there is a risk

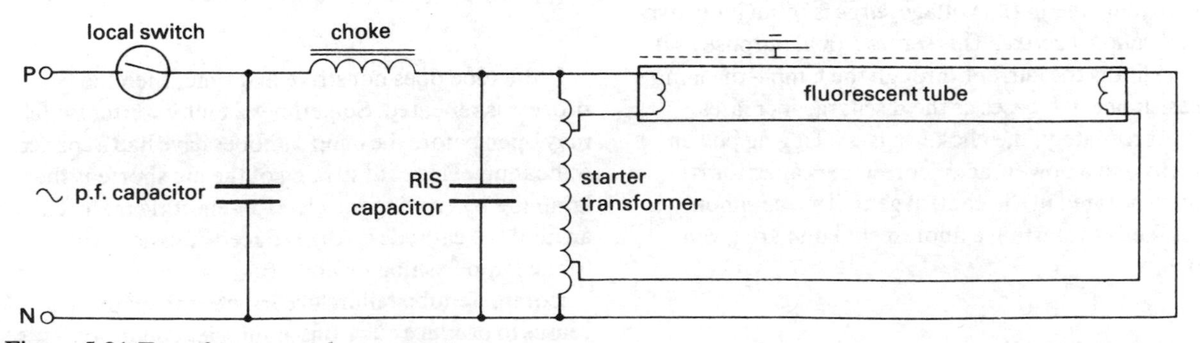

Figure 5.21 *Transformer quickstart circuit*

* It should be pointed out that conventional starters may eventually be replaced by **electronic igniters**, which have a much greater life expectancy.

that rotating parts may appear to be stationary. This phenomenon is called 'stroboscopic effect'. It occurs in discharge lamps because their discharge is being extinguished twice every cycle, which causes them to flicker every 10 ms. This does not happen with incandescent lamps because their filaments do not cool down fast enough to show any signs of the cyclic supply variation. Perhaps the simplest way to understand stroboscopic effect is to consider the spokes of a rotating wheel. At the moment in time when the discharge lamp is receiving zero voltage, a spoke is always in the position that was occupied by another spoke whose particular time difference is equal to one half-cycle of the supply frequency.

To overcome the problem of the stroboscopic effect, twin lamp fittings can be used and wired as a *lead-lag* circuit such that current entering one lamp is artificially *phase-displaced* with current entering the other lamp. This results in both lamps being extinguished at different time intervals in the a.c. cycle. A typical lead-lag circuit is shown in Figure 5.22. An alternative arrangement when a three-phase supply is available is to wire adjacent lamps to different phases, e.g. red, yellow, blue, red, yellow, blue, etc. In this way, one lamp will extinguish every 120° (elect). A further solution to the problem is to use a mixture of incandescent and discharge lighting.

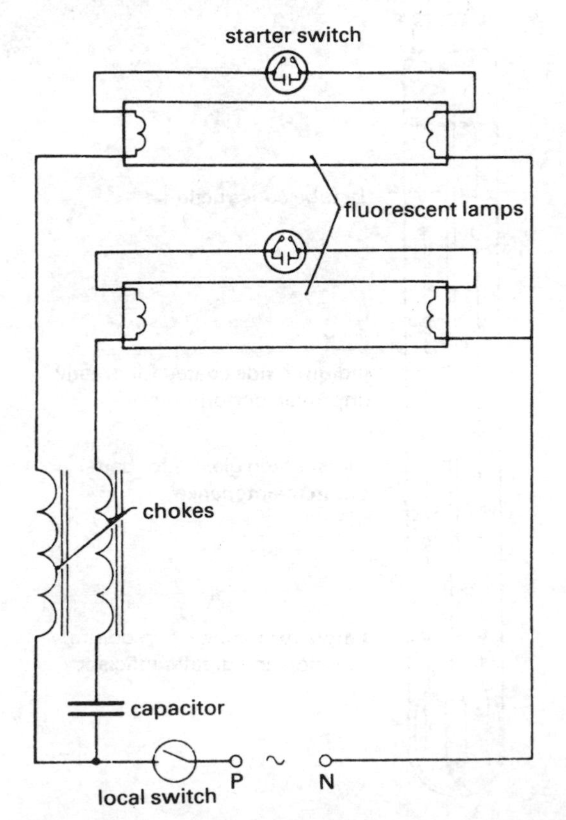

Figure 5.22 *Typical lead-lag circuit*

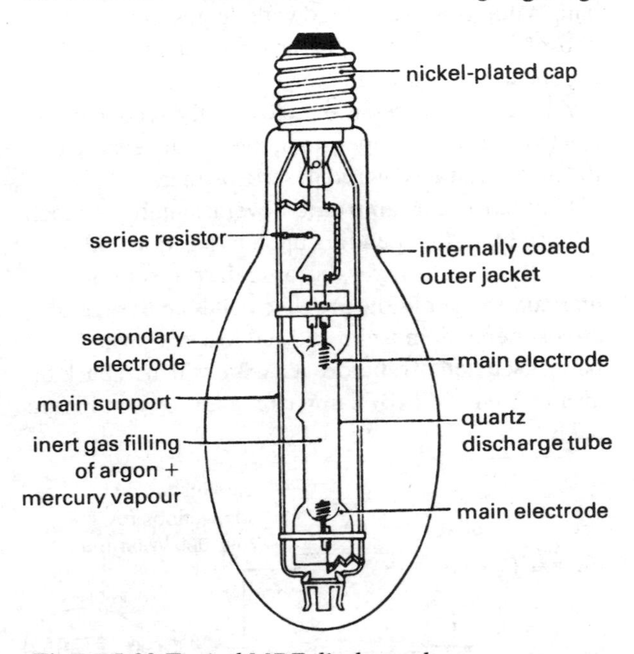

Figure 5.23 *Typical MBF discharge lamp*

High-pressure mercury fluorescent lamps

Figure 5.23 shows a typical high-pressure mercury vapour lamp. This particular lamp is classified MBF, having a quartz envelope loaded below 100 W/cm of arc length. It is suitable for operating in any position and therefore may be listed as an MBF/U lamp (U standing for universal operating position).

The lamp is found in several sizes, ranging from 50 to 1000 W (230–250 V range). Lamps are fitted with Edison screw and Goliath Edison screw lamp caps. They have lighting design lumens ranging from 1800 to 58 000 lm, with an efficacy range of 38–56 lm/W. Lamp life is ~ 7500 h.

Ionisation inside the quartz arc tube is achieved by means of a secondary electrode placed close to one of the main electrodes. In series with the secondary electrode is a high resistance, which is connected to the other main electrode. When the lamp is switched on a discharge occurs between the main electrode and the secondary electrode. This ionises the inert argon gas filling and warms up the discharge tube, allowing an arc to strike between the two main electrodes. The mercury condensed on the inside walls of the discharge tube now becomes vaporised and takes over the discharge. The inside surface of the lamp's outer bulb is coated with fluorescent phosphors, which convert the ultraviolet radiation produced in the arc tube to visible light.

As the lamp's pressure builds up, the secondary electrode ceases to operate in the circuit because of the high resistance wired in series with it.

In practice, the lamp takes several minutes to reach full brightness, depending upon its location and type of enclosure. However, because the pressure of the mercury vapour in the arc tube builds up to several atmospheres, it must be allowed to cool down after being switched off, before it can be switched back on again. A typical MBF lamp circuit is shown in Figure 5.24.

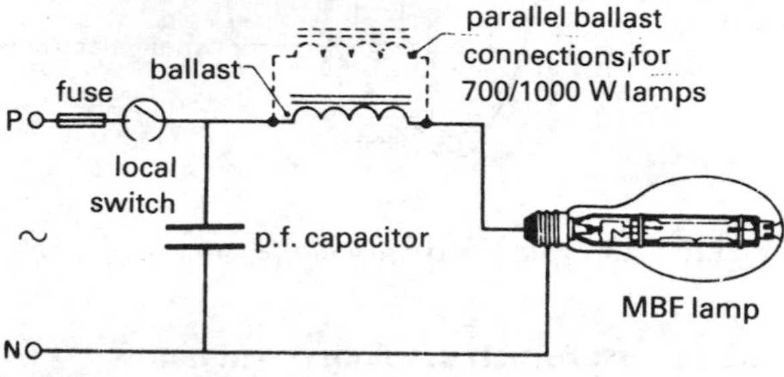

Figure 5.24 *Wiring circuit for MBF lamp*

Mercury discharge lamps produce *line spectra*, i.e. they emit energy at certain wavelengths only—they are deficient in the colour red, the strongest lines being purple, green, yellow and ultraviolet. The use of fluorescent powders, particularly a red fluorescing phosphor, greatly improves the colour rendering and efficacy of mercury discharge lamps. The lamps have wide application, particularly in street lighting, industrial lighting, showrooms, shops, etc.

A lamp that does not light may be the result of an open circuit in the wiring or the ballast, incorrect circuit connections, end of lamp life, or even not giving the lamp sufficient time to restrike. Low voltage or incorrect ballast tapping could result in poor light output. Checks should be made to see if the voltage is correct and to ensure that there are no signs of the ballast overheating.

Low-pressure sodium vapour lamps

This lamp, classified as SOX, is shown in Figure 5.25. It consists of a U-shaped arc tube containing metallic sodium and an inert gas such as neon. The dimple formation provides cool spots to retain the metallic sodium, thus preventing mirroring. The discharge tube is enclosed in a tubular bulb whose inside surface has a reflector coating of indium oxide. This coating acts as an infrared reflector, keeping the discharge tube temperature at an optimum for minimum input power, thereby maintaining full efficacy.

Five common lamp wattages are 35, 55, 90, 135 and 180 W. The range of lighting design lumens is between 4300 and 31 500 lumens, giving efficacies of

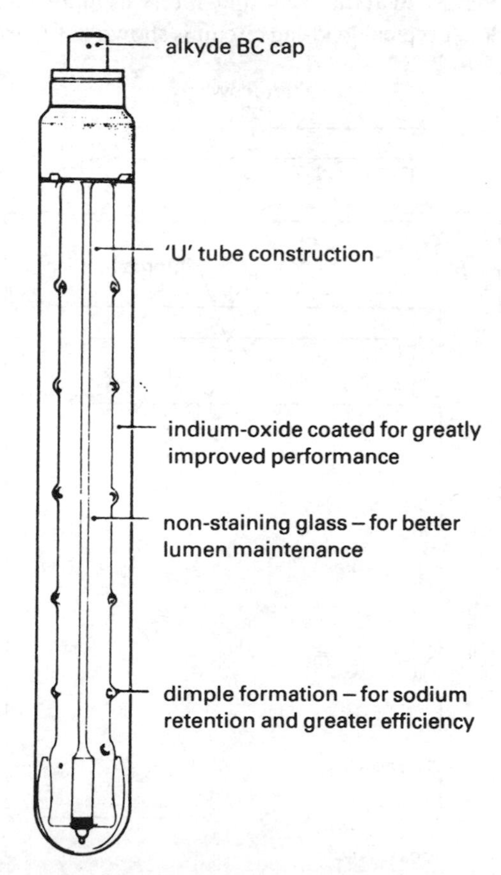

Figure 5.25 *Typical SOX low-pressure sodium lamp*

123 and 175 lm/W, respectively. Burning position is in the horizontal ±20°, but the 35 W and 55 W lamps are able to operate in the cap-up position. The lamps are made to operate on 230–250 V supplies and their rated life is approximately 6000 h.

The low-pressure sodium lamp has a much longer arc length than the previously mentioned mercury lamp and it therefore requires a higher voltage to start it (between 480 and 650 V). This is achieved by means of a leakage transformer, which also acts later as a current limiting device. The smaller rated lamps, namely the 35 W and 55 W lamps are often started by means of an external igniter.

When the SOX lamp is connected in circuit to the a.c. supply, the discharge is first struck in the neon gas, giving off a characteristic red glow. The lamp normally takes between 6 and 15 min to reach full brightness. As it warms up, the metallic sodium condensed on the walls of the inner tube begins to evaporate, eventually taking over the neon discharge. The colour of the lamp changes from red to monochromatic yellow. The lamp, therefore, has a poor colour rendering since all surrounding colours appear to look yellow. Despite this disadvantage, the lamp has a high efficacy and is used extensively for street lighting and floodlighting and in situations where colour discrimination is not important. A typical circuit diagram of the lamp is shown in Figure 5.26.

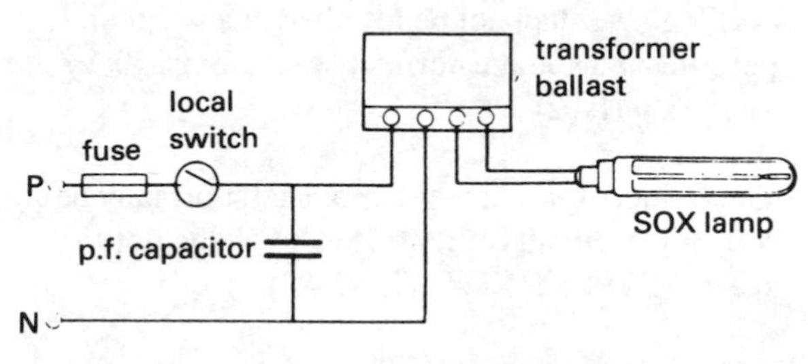

Figure 5.26 *Wiring circuit for SOX lamp*

A lamp that does not light should first be replaced with a sound one, although it is possible to have a circuit fault either in the actual wiring or ballast. One should check the supply voltage, particularly if the lamp is working but showing poor light output. Poor light output could also stem from a lamp failing or wrong ballast tapping.

High-pressure sodium vapour lamps

A typical high-pressure sodium lamp is shown in Figure 5.27. The type is classified as SON but there are two other versions, namely SON-T (a tubular clear glass type) and SON-TD (a tubular double-ended clear quartz). The SON lamp consists of an arc tube made of sintered aluminium oxide capable of withstanding intense chemical activity at high temperature. It has an internally coated isothermal, outer envelope, giving a wide spread of illumination in a golden-white light. A circuit diagram of the lamp is shown in Figure 5.28.

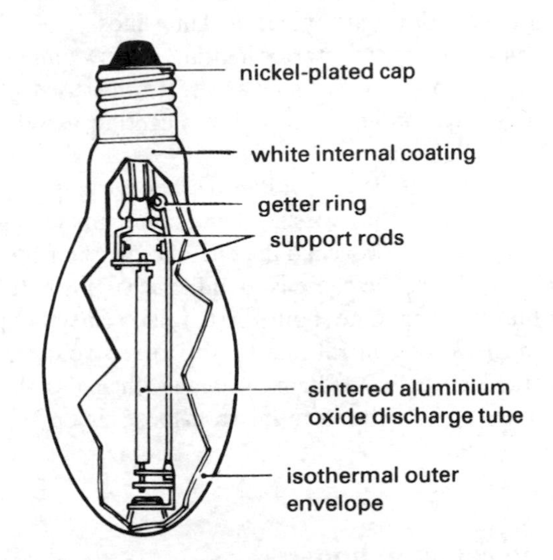

Figure 5.27 *SON high-pressure sodium lamp*

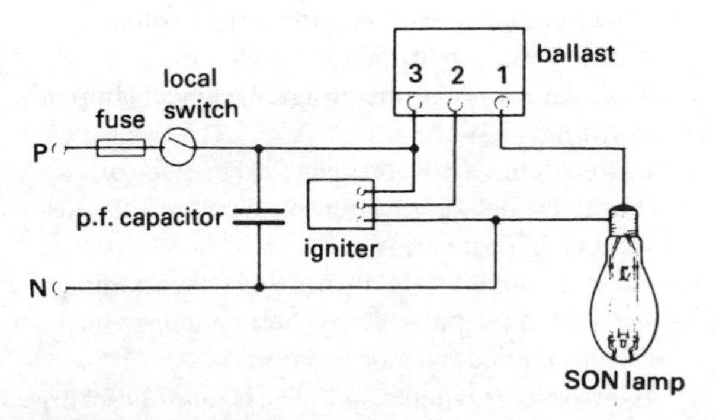

Figure 5.28 *Wiring for SON lamp*

The lamp is started by a high-voltage pulse applied by an igniter, which ceases to function once the arc has struck. It takes approximately 5 min to reach full brightness and it will normally restrike within 1 min of extinction. Wattage ranges for the SON lamp are between 70 and 1000 W, with respective lighting design lumens between 5300 and 110 000 lm. The lamp is usually fitted with a GES lamp cap and may be used in any position. Its rated life is approximately 6000 h and it operates from a 230–250 V supply. Lamp efficacy range is between 82 and 120 lm/W and it has reasonable colour rendering properties, with blues and greens a little subdued and reds and yellows somewhat enhanced. Because all surrounding colours are distinguishable, the lamp finds considerable use in interior lighting, e.g. swimming pools and modern shopping areas. It is also used extensively in industry for high-bay lighting, as well as in street lighting.

Lamp problems are similar to other discharge lamps. A lamp that does not light, although looking intact, may have reached the end of its working life, but sometimes the cause is found to be supply failure or faulty wiring. One should first of all try a new lamp, then check the control gear for the correct voltage, particularly ballast voltage. A lamp with an unstable light output may be suffering from loose circuit connections.

General lamp hints

1. Always isolate the circuit before inserting or replacing a lamp.
2. Check lamp circuit fuse for suitability.
3. Check replacement lamp for correct voltage, wattage and cap fitting.
4. Take care when inserting a replacement lamp in its fitting.
5. Protect lamp against mechanical damage or thermal shock, e.g. water splashing on bulb; also protect against vibration.
6. Read manufacturers' instructions on installing special lamps, particularly their handling and recommended burning position.
7. Exercise care when disposing of lamps. Discharge tubes should be broken in a container, outdoors. One should not attempt to break the inner arc tubes of high-pressure lamps. Caution is needed with low-pressure sodium lamps because of the risks of fire and flying glass.
8. Exercise care when handling control gear, such as transformer ballasts. Such apparatus is quite heavy and could cause personal injury if accidentally dropped on hands or feet.
9. Be aware of the high voltage present in ballast units and take care when handling capacitor units in case they are not fitted with discharge resistors to de-energise them. Seek manufacturers' instructions before disposal of control gear, particularly capacitors containing chlorinated diphenol impregnants—they may constitute an environmental hazard.

Regulation requirements

The IEE Wiring Regulations incorporate a number of requirements concerning lighting installations. Some of these are:

All filament lamps shall be so placed or so guarded as to prevent ignition of any flammable materials. Any shade or guard used for this purpose shall be suitable to withstand the heat from the lamp. (Reg. 422–01)

A ceiling rose shall not be installed in any circuit operating at a voltage normally exceeding 250 V. (Reg. 553–04–02)

Lampholders for filament lamps shall normally be used only in circuits operating at a voltage not exceeding 250 V. (Reg. 553–03–02)

Where centre-contact bayonet or Edison-type screw lampholders are connected to a source of supply having an earthed neutral conductor, the outer or screwed contact shall be connected to that conductor. Where such lampholders are used in circuits having neither pole connected with earth, in any damp situation or in any situation in which they can readily be touched by a person in contact with or standing on earthed metal, they shall be provided with a protective shield of insulating material complying with Appendix B to BS 98, or shall be placed or so safeguarded, so that neither the lamp cap nor the outer or screwed contact of the lampholder can be inadvertently touched when the lamp cap is engaged with that contact. (25). (See also the HSE Advisory Publication, SHW 1911.)

Consideration should be given to: the avoidance of mechanical strain on any cable termination inside the ceiling rose or similar accessory; PVC outlet boxes should not be used where the temperature exceeds 60°C and they should not support luminaires of more than 3 kg in mass; sizes of flexible cords supporting luminaires on their own are restricted to masses of 2 kg for 0.5 mm^2, 3 kg for 0.75 mm^2 and 5 kg for 1.0 mm^2; flexible cords joining ceiling rose to lampholder should be suitable for an operating temperature of not less than 85°C; earthing terminals are required at lighting points and switch points; and control gear capacitors should be provided with a means of discharge immediately the supply is disconnected.

With regard to discharge lighting, a number of regulations also apply, such as current rating of switches, loading of circuits, control gear, screening of live parts, as well as maximum voltage permissible on circuits operating above low voltage. Requirements concerning neon sign installations are also provided.

Concerning switches, it is known that transient charging currents taken by capacitors can be up to 20 times greater than the normal running current. When a circuit is switched off the collapsing magnetic field of a choke may cause a transient voltage surge of several thousand volts. Switches should be rated for twice the steady load current and ideally the switch should be of the type that provides a quick-make and slow-break action.

On the loading of circuits, it is recommended that the voltamperes of the circuit be determined by multiplying the rated lamp power (watts) by a factor of 1.8. This multiplier takes into consideration a circuit power factor of not less than 0.85, and also control gear losses and harmonic currents. Harmonic currents derive from several sources, but in discharge lamps they are mainly produced by the magnetic circuit of lamp ballasts, e.g. the inductive chokes, and result in distortion of the normal a.c. waveform.

Unfortunately, correcting the discharge lamp's poor power factor, to reduce the circuit current, increases the percentage of harmonic current. It is for this reason that full-size neutrals should be used in three-phase, four-wire systems since harmonic currents are additive in the neutral. Minimum neutral current flows when all three phases are balanced, and maximum current flows when there is no load on any one phase and full load on the other two.

To determine the cable size for a particular lamp, say a 40 W fluorescent tube in a switch start circuit, the procedure for finding circuit current is as follows:

$$I = \frac{P}{V} = \frac{40 \times 1.8}{240} = 0.3 \text{ A.}$$

When the lamp circuit details are given, e.g. control gear losses 10 W, power factor 0.85 lagging and harmonic current 0.042 A, then the current required is found from

$$I = \frac{P}{V \times \text{p.f.}} \text{ where } P = 40 + 10 \text{ W}$$

$$= \frac{50}{240 \times 0.85}$$

$$= 0.245 \text{ A.}$$

By adding the harmonic current to this, the total circuit current is:

$$I = 0.245 + 0.042$$
$$= 0.287 \text{ A.}$$

This figure compares favourably with the one above, using the 1.8 multiplier.

High-voltage installations

Several regulations relate to discharge lighting installations using voltages between 600 and 5000 V to earth. Besides not exceeding 5000 V to earth, circuits supplied from a transformer having a rating above 500 W (input) shall be provided with the means for automatic disconnection of the supply in the event of a short circuit or earth leakage current that exceeds 20% of the normal steady current in the circuit. Ancilliary equipment such as inductors and capacitors are to be either totally enclosed in a substantial, earthed, metal container, or alternatively placed in a suitably ventilated enclosure of non-combustible material or fire-resisting construction. A notice, DANGER—HIGH VOLTAGE, must be placed and maintained on every container or enclosure that is accessible to unauthorised persons.

High-voltage discharge lamps are generally supplied by double-wound transformers, and autotransformers may be used on two-wire circuits that do not exceed 1.5 kW measured on open circuit—one pole of the supply must be connected to earth and the control switch must be of the double-pole type. Circuit conductors may be isolated through an *interlock*, in addition to the switch normally used for controlling the circuit; *local isolation*, by means of a plug and socket outlet (or other effective means) in addition to the normal control switch; *switch with removable handle*, switch that can be locked or distribution board that can be locked. Where more than one such switch or distribution board is installed, all keys and removable handles shall be non-interchangeable.

A fireman's emergency switch is required for all exterior installations and unattended interior installations where they are used for window-display lighting. The switch must be capable of isolating all live conductors and it must be coloured red and marked FIREMAN'S SWITCH (with regulation lettering). It must be clearly marked ON and OFF, with the *off* position at the top; it should be fixed in a conspicuous position, reasonably accessible to firemen; and, except in the case of an agreement to the contrary with the local fire brigade authority, it should be placed at not more than 2.75 m from the ground. A typical control circuit for a neon sign is shown in Figure 5.29.

Other requirements specify the positioning of the fireman's switch, preferably near the sign in exterior installations or by the main entrance (or in a position agreed with the fire brigade authority) in interior installations. Sections of discharge lighting may be controlled by one fireman's switch but where more than one switch is installed on any one building, they should be clearly marked to indicate the installation or section that they control—the local fire brigade should be notified of the arrangements.

High-voltage cables should be armoured or metal sheathed, except for short connections between the sign tubes. They shall be adequately supported and suitably protected. The high-voltage cables shall be distinguished by tabs or labels marked DANGER, securely attached at intervals not greater than 1.5 m and regulation size lettering must be used. The earthed return conductor from an electrode to the transformer terminal must not be less than 2.5 mm^2.

Emergency lighting

The Fire Precautions Act 1971 and the Health and Safety Act 1974 contain statutory requirements in all places of work and public resort, to provide adequate means of escape. BS 5266 is a Code of Practice dealing with emergency lighting of premises, laying down minimum standards for the design, implementation and certification of installations.

Emergency lighting is lighting provided for use when the normal lighting fails. *Escape lighting* is part of the emergency lighting that is provided to ensure that the means of escape can be safely and effectively used at all material times. A *maintained emergency lighting system* is one where all emergency lighting lamps are in operation at *all* material times, whereas a *non-maintained emergency lighting system* is one where all emergency lighting lamps are in operation *only* when the normal lighting fails. An emergency lighting fitting or luminaire containing two lamps, one energised from the normal lighting supply and the other from an emergency lighting supply, is called a *sustained luminaire*.

The function of escape lighting is to:

(a) indicate clearly and unambiguously the escape routes,
(b) provide illumination along such routes to allow safe movement towards and through the exits provided,

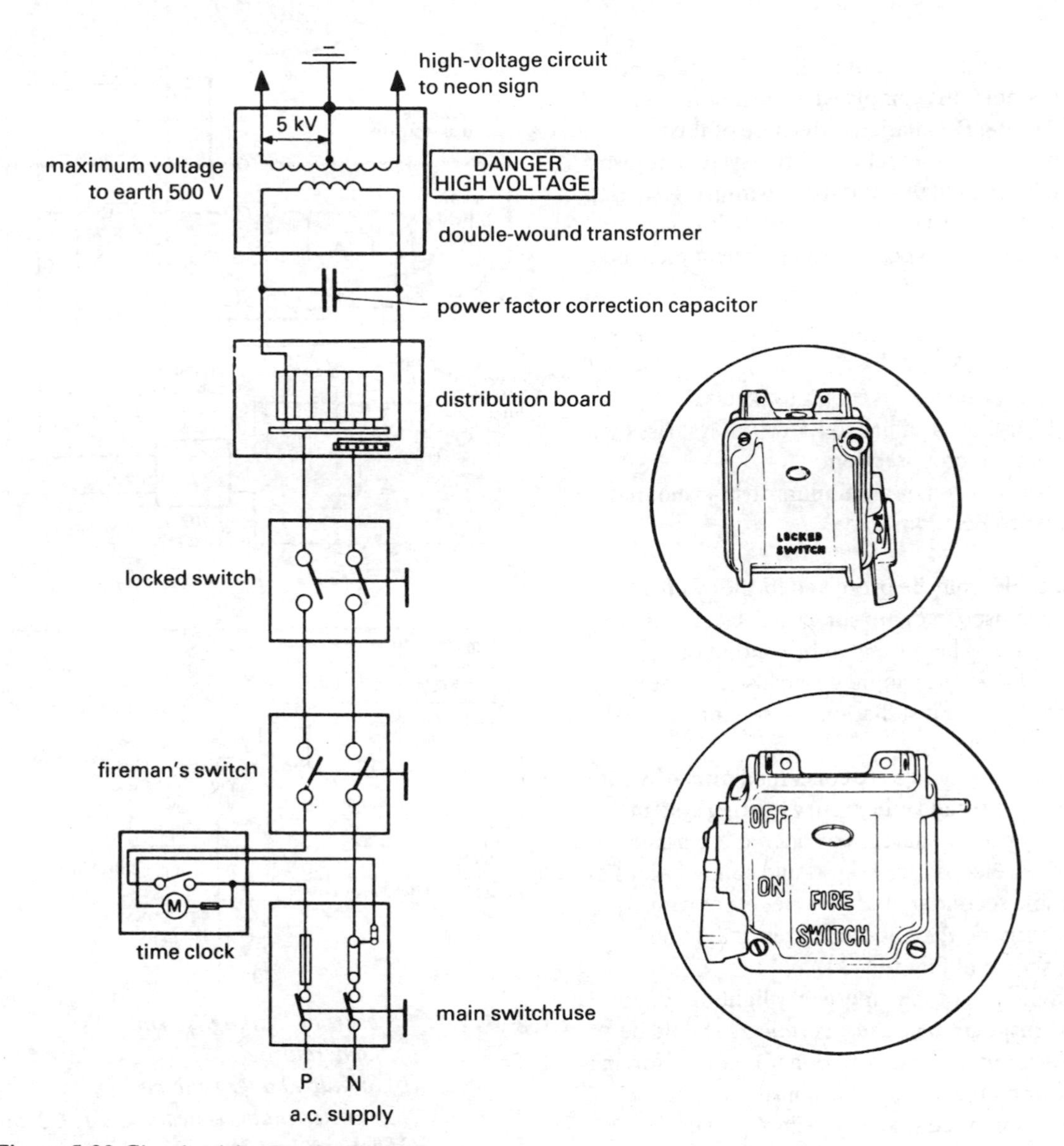

Figure 5.29 *Circuit wiring to a neon sign*

(c) ensure that fire-alarm call points and fire-fighting equipment provided along escape routes can be readily located.

Escape lighting luminaires should be sited near each exit door and emergency exit door and also at points where it is necessary to emphasise the position of potential hazards, e.g.

(a) near each intersection of corridors,
(b) near each change of direction (other than on a staircase),
(c) near each staircase so that each flight of stairs receives direct light,
(d) near any other change of floor level that may constitute a hazard,
(e) outside each final exit and close to it.

The BS 5266 Codes of Practice recommends that escape route illuminance should be not less than 0.2 lux and should be provided within 5 s of failure of the normal lighting. (This may be extended to 15 s at the discretion of the enforcing authority.)

From the point of view of wiring, the emergency lighting system must comply with the requirements of the IEE Wiring Regulations. Because of the importance of the emergency lighting system, in terms of its reliability, a high standard of wiring is essential. Cable selection must consider the problem associated with voltage drop. Types of cable(s) recommended include:

(a) m.i.m.s. cables (mineral-insulated metal sheath cables) to BS 6207,
(b) elastomer-insulated cables to BS 6007,
(c) PVC-insulated cables to BS 6004—not less than 1.0 mm^2 in cross-section,
(d) wire-armoured or aluminium-strip-armoured cables to BS 6346.

M.i.m.s. cables may be protected further with a PVC sheath when used in damp situations. Cables under (b) and (c) are to be protected by conduit or trunking. The wiring should be exclusive to the emergency lighting installation and separate from the wiring of any other circuit.

The Codes of Practice are concerned primarily with permanently installed emergency lighting systems, and the only power sources considered are prime motor-driven electric generators and combinations of rechargeable secondary batteries together with suitable chargers. A typical emergency lighting system is shown in Figure 5.30.

On completion of the emergency lighting system wiring, an inspection and test certificate should be given by a competent person, either by the enforcing authority or by the installer responsible.

It is recommended that a system of preventive maintenance be programmed and that routine testing be carried out as follows: central battery systems—twice yearly, self-contained luminaires and internally illuminated signs—once a month; engine-driven plant (with or without standby batteries)—once a month.

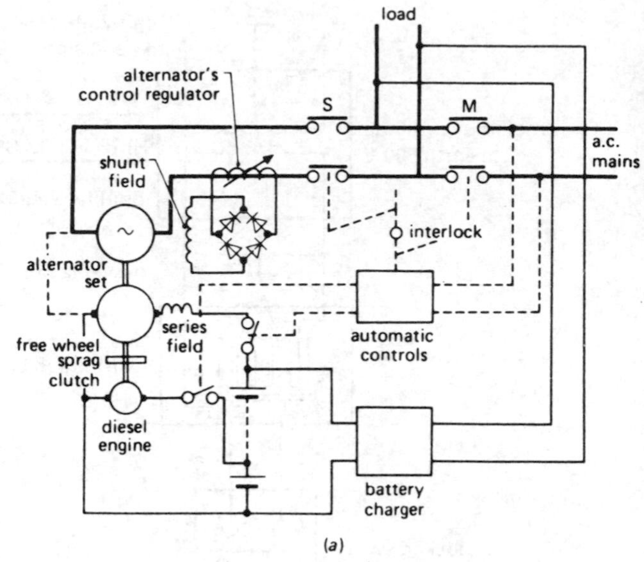

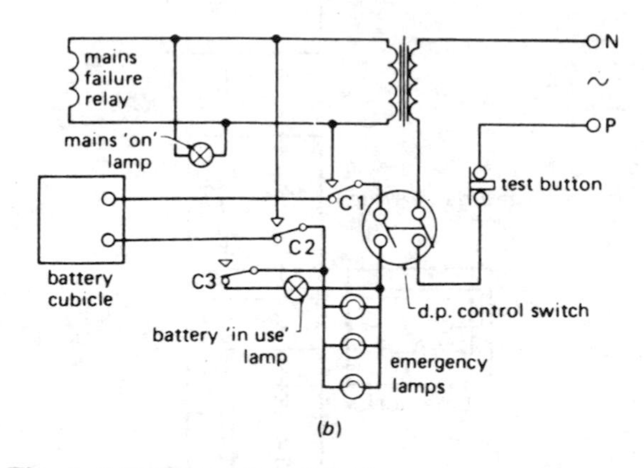

Figure 5.30 *Emergency lighting systems*
(a) Generator set
(b) Standby battery cubicle
Note: when mains is healthy, C1, C2 and C3 contacts are lifted by mains failure relay.

Lumen method calculations

The method of calculating illuminance described as the *point by point method* is found in Volume 2. The lumen method considers the whole lamp output in lumens or, more accurately, the lighting design lumens (LDL) quoted earlier in the chapter. This method will only give the average illuminance on the horizontal plane; it will not give the actual value at any one point.

In practice, the problem is finding out how many luminaires are required for a particular room. Recommended standard service illuminance values have been compiled by the Illuminating Engineering

Society (IES). The values shown below are typical for a range of various tasks.

Task group and typical task or interior	*Standard service illuminance* (lux)
Storage areas and plant rooms with no continuous work	150
Casual work	200
Rough work: rough machining and assembly	300
Routine work: offices, control rooms, medium machining and assembly	500
Demanding work: deep-plan, drawing or business machine offices; inspection of medium machining	750
Fine work: hand engraving	1000
Minute work: inspection of very fine assembly	3000

The lumen method uses the following formula:

$$N = \frac{E \times A}{F \times C \times M}$$

where N is the number of luminaires required
E is the average illuminance required
A is the area requiring to be lit
F is the lumen output of the lamp(s)
C is the coefficient of utilisation
M is the maintenance factor.

The coefficient of utilisation allows for the losses incurred by the absorption of light by walls, ceiling and floor, type of lamp fitting or reflector. It may be thought of as the ratio of the total flux reaching the working plane to the total flux, i.e.

$$C = \frac{\text{Useful flux}}{\text{Total emitted flux}}$$

The ratio is always less than unity.

The maintenance factor, again, can be thought of as a ratio of the illumination from a dirty installation to that from the same installation when clean. It is normally around 0.8 and allows for losses due to the accumulation of dirt and the ageing of luminaires.

Where it is simply required to determine the total number of lumens for a particular room, then:

$$\text{Lamp flux } (F) = \frac{E \times A}{C \times M} \text{ lumens.}$$

Example 1

An office measuring 10 m × 8 m requires to be illuminated to a level of 500 lux. Calculate the total lamp flux if the utilisation and maintenance factors are 0.9 and 0.8, respectively.

$$\text{Lamp flux } (F) = \frac{500 \times 80}{0.9 \times 0.8}$$

$$= 55\,556 \text{ lumens.}$$

If the lamps to be used are fluorescent tubes having an efficacy of 52 lm/W, determine the total lamp power required.

$$\text{Total lamp power } (P) = \frac{55\,556}{52}$$

$$= 1068 \text{ W.}$$

Example 2

A classroom measuring 16 m by 16 m requires an illuminance at desk level to be 350 lux. The mounting height of the lamps above desk level is 2 m. Calculate the number of 80 W fluorescent lamps required, assuming the lighting design lumens to be 5100, coefficient of utilisation 0.6 and maintenance factor 0.8.

Total number of lamps required

$$N = \frac{E \times A}{F \times C \times M}$$

$$= \frac{350 \times 16 \times 16}{5100 \times 0.6 \times 0.8}$$

$$= 36.6$$

This figure is approaching 37 and it may be justifiable to install 38 lamps.

Spacing/mounting-height ratio

If the spacing between luminaires is too wide, there will be a serious drop in illuminance in the areas between them. It is recommended that the illuminance on the working plane should not fall below 70% of that which is directly below the luminaire. The ratio of the spacing between adjacent luminaires to the mounting height above the working plane is called the spacing-height ratio (*S*/*H*). In practice, the variation of spacing to mounting is from 1:1 to 2:1 and the working plane is regarded as 0.85 m above the floor.

Measuring light

Very often, an *illuminance meter* is used to check that the achieved illuminance at various points in a lighting installation agrees with a calculated or specified value. If readings have to be taken during daylight, a rough allowance can be made by subtracting the readings taken with the lights off from the readings taken with the lights on, although it must be emphasised again that readings will not be accurate, particularly if the daylight is variable. To find the average illuminance, a number of readings at evenly spaced points in the room should be taken. An ideal procedure is to divide the plan area into a number of rectangles and take readings at the centre of each rectangle. If these points are marked or identified with sticky tape, then the light meter can be held above these points at the required level to obtain an average illuminance.

Generally speaking, there are two types of light meter: one where the meter incorporates a precalibrated luminance that has to be varied until it matches the luminance to be measured; the other where the area whose luminance is to be measured is viewed through the meter and a reading is taken directly from the scale.

A typical light meter is shown in Figure 5.31. It operates from a *photocell* with an appropriate diffuser set into a specially designed well to overcome any errors caused by the angle of incidence of the light rays on the photocell. A spectrum-correcting filter is also fitted, which corrects the spectral response of the cell to provide a direct reading even for mercury and sodium lamps. In addition, the photocell is mounted such that no shadow that would affect the accuracy of the reading will be cast on the cell during measurement. The light meter is calibrated in lux and provides two switch-selected ranges for measuring lighting intensities up to 2000 lux.

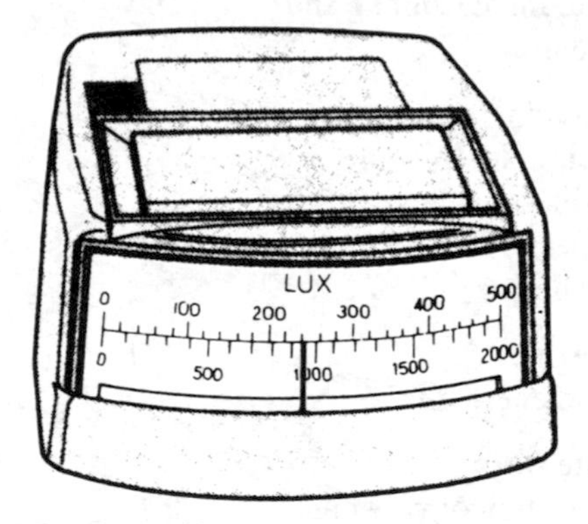

Figure 5.31 *Typical light meter*

A photocell consists of an iron back-plate that supports a thin layer of selenium upon which is coated a thin transparent layer of gold to allow light to penetrate the gold-selenium boundary. This results in the liberation of electrons, passing from the gold to the selenium. Connections are made, between the back-plate and the gold layer, to a microammeter instrument calibrated in lux. This allows current to flow from the iron plate to the gold and through the meter.

Motor installations

An electric motor is a machine that converts electrical energy into mechanical energy. It has the job of providing a driving torque for some specific task or operation. The selection of a particular type of motor may involve a number of factors, such as:

(a) rated output
(b) rated voltage
(c) type of supply
(d) type of enclosure
(e) method of starting
(f) speed control

The above list is by no means complete. A motor's rated output is expressed in watts or kilowatts and its size has to satisfy the type of mechanical load connected to its drive shaft. It is not often appreciated that a motor is in fact capable of developing more torque than its actual output rating. One question that might be asked is whether the motor will be required to transmit a constant torque over its entire speed range or whether it will need to do this for certain speeds only. In some cases it may be necessary to consider a motor's *duty cycle* or *duty factor*, which considers the motor getting too hot if it is started too frequently. Some motors have a *continuous duty rating*, which allows them to develop their rated output continously without exceeding safe temperature limits. It should be pointed out that motors can take in-rush starting currents up to eight times their normal running current. This factor has not escaped the IEE Regulations, Reg. 331–01 concerning the harmful effects this might have on other equipment or other services.

The rated voltage is the voltage between line terminals of the motor at its rated output. A motor is normally suitable only for the voltage marked on its nameplate and for alternating current (a.c.) supplies it is 240 V (single-phase) and 415 V (three-phase). For direct current (d.c.) supplies it is between 100 and 250 V. There may, however, be variations to these voltages for special machines, such as 110 V a.c. for experimental work as well as for safety.

With regard to the type of supply, the choice is whether or not the motor is going to be operating from an a.c. or a d.c. source. There are numerous types of single-phase a.c. motor, such as the split-phase cage induction motor, shaded-pole motor and synchronous motor. These tend to be fractional output motors which find considerable use in domestic premises, e.g. washing machines, hand mixers, clocks and timers. The motor used in a vacuum cleaner is a d.c. series motor adapted for a.c. use, called a *universal motor*. On three-phase supplies, the most common motor is the *three-phase cage induction motor* but there are other types, like the *synchronous motor* and *a.c. commutator motor*. Motors requiring a d.c. supply are all commutator motors, such as the *series*, *shunt* and *compound motors*, named according to their field winding arrangement. These motors, like the three-phase a.c. motors, have a wide industrial use.

A motor and its controlgear may require protection against dusty, humid, corrosive or even explosive atmospheres. To safeguard against these conditions, several different types of enclosure are made. For example, in an environment where there is a risk of dripping liquids, a *drip-proof motor* is suitable. It will be noticed that this motor has openings at either end, to give it greater ventilation and to help stop the casing from overheating. The openings are fitted with a wire-mesh screen to prevent objects from entering the working parts and causing damage. In places which are dusty or where there is a risk of splashing, then a *totally enclosed motor* would be suitable. This type of motor is shown in Figure 5.32. Its enclosure consists of a number of moulded fins spread around the outer casing, which serve the purpose of increasing the overall surface area to allow heat to escape more freely. These motors are not completely air-tight but they are often fitted with internal fans to facilitate cooling. Other types found are screen-protected, pipe-ventilated and flameproof. It is worth

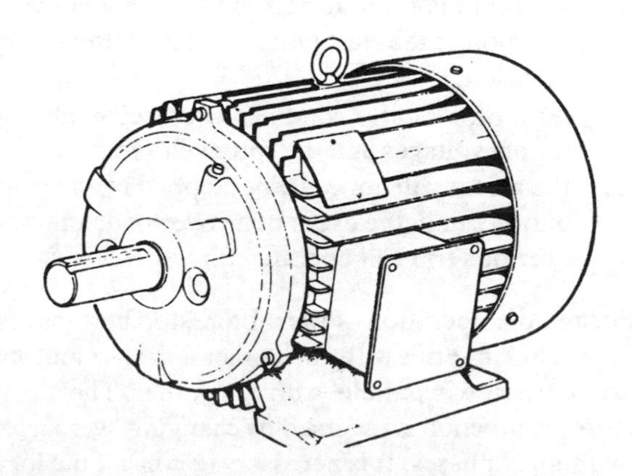

Figure 5.32 *Totally enclosed motor*

mentioning at this stage that motors and their associated loads can make a considerable noise when rotating, owing to the circulating air inside their casing as well as bearing noise and that attributable to electromagnetic excitation. For this reason, motors are sometimes sited away from the work place.

As mentioned earlier, a motor can take a high starting current and this can cause damage to its windings or cause a disturbance on the supply system, such as a voltage drop. Reference should be made to Section 552 of the IEE Regulations regarding the requirements for rotating machines. The majority of small motors can be connected directly to the incoming supply with little effect, but the larger motors are often started by some arrangement to reduce their terminal voltage (since current depends on voltage). This is achieved in several ways, e.g. star-delta starting, autotransformer starting or rotor-resistance starting, as in the case of a.c. connected motors; resistance starters or contactor panels, as in the case of d.c. motors. Other arrangements might involve electronic control or some method of reducing the initial supply voltage. Some of the more common methods will now be discussed.

Figure 5.33 shows a typical *direct-on-line contactor starter* for a three-phase induction motor. Its operation is as follows:

(a) the main isolator will be closed,
(b) the start button will be pressed and a current will take the route shown by the arrows,
(c) the contactor will become energised by the coil and close,
(d) the retaining contact (S) will supplement the function of the start button and the latter can be released,
(e) the motor's stator windings will receive full supply voltages and the rotor will run,
(f) the motor will normally be stopped by the stop button but in the event of an overload, an overload trip will operate.

For remote operation of the motor, stop buttons are connected in series with each other and start buttons are connected in parallel with each other. The motor's direction is reversed by changing over any two supply phases. It is necessary to consult the local REC when starting large induction motors.

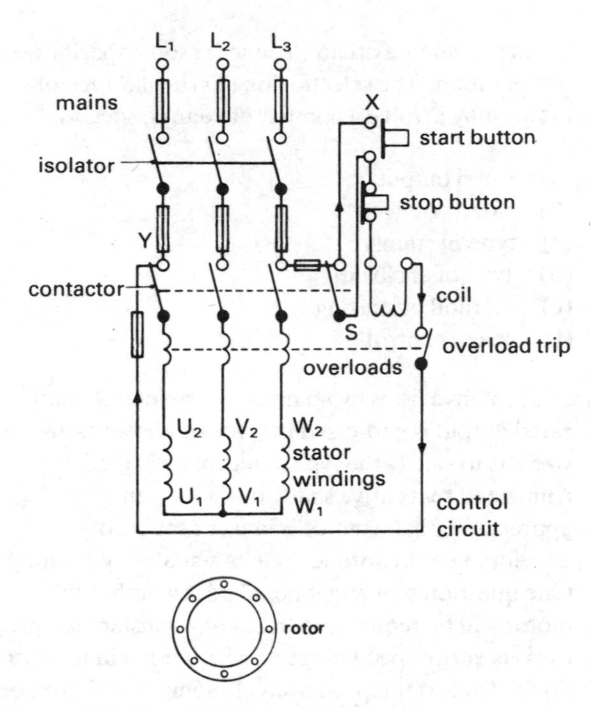

Figure 5.33 *Simple direct-on-line starter*

Figure 5.34(a) shows the connections of a simple hand-operated *star-delta starter*; once the isolator has been closed, the changeover switch has to be moved from the star (start) position to the delta (run) position after several seconds. Once in this latter position, the changeover switch is locked in by an internal catch mechanism, which is automatically disengaged if an overload occurs. Motors requiring this method of starting need to have six stator terminals. This method of starting provides a reduced voltage across each phase winding (58%) and means that the line current at starting is only about one-third of the value it would normally be if it were switched directly on to the supply. Unfortunately, because the induction motor has a relatively poor starting torque, reducing the motor's voltage does not help since the starting torque is proportional to the voltage squared and therefore the method is limited to those motors not required to be started on heavy loads.

A third method of starting three-phase induction motors is by using an *autotransformer*. This method is ofen used when the previous method is not suitable, either because the starting torque is too low or because the motor has only three stator terminals

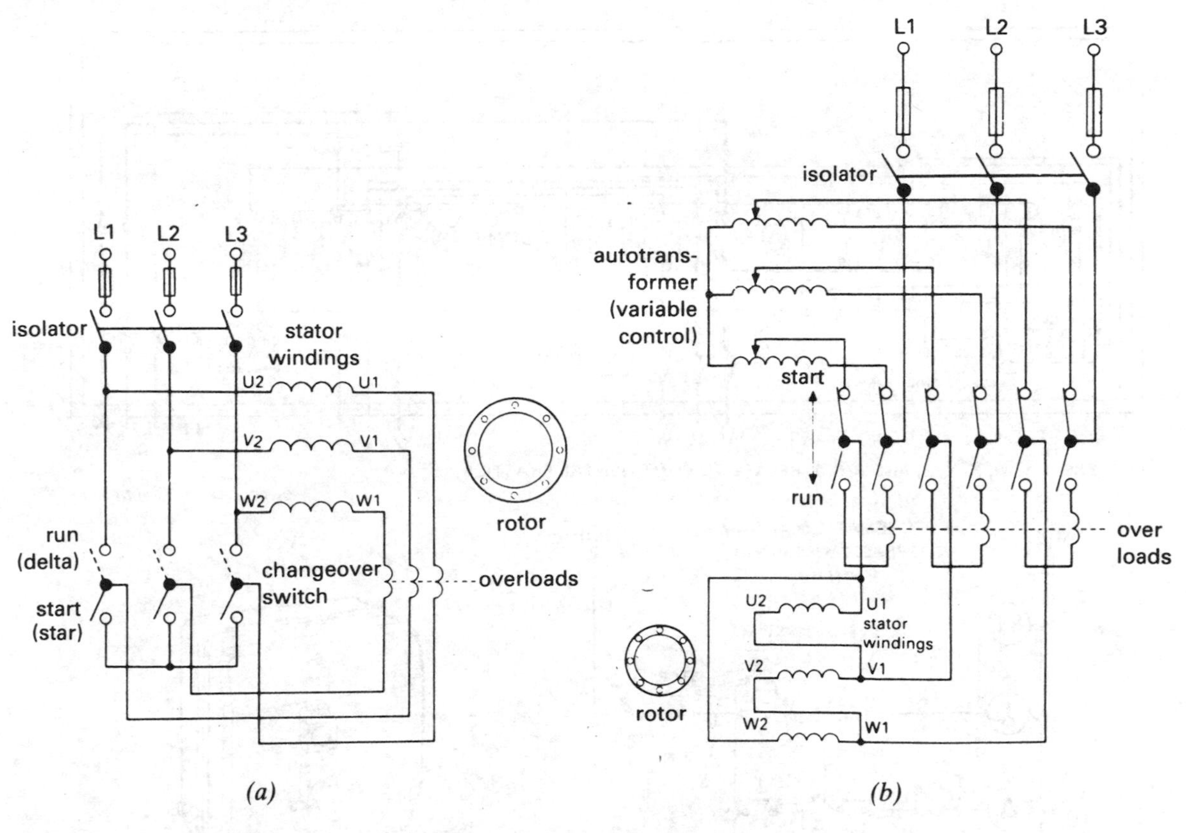

Figure 5.34 *Simple starting methods for three-phase cage induction motors*
(a) Star-delta starter
(b) Autotransformer
Note: circuit fuses have been omitted

brought out for connection. Briefly, it will be seen that the autotransformer is used as a variable control for the start position, depending on what torque is required. For example, 40% supply tapping would be suitable for a motor's starting torque of 16% whereas at 75% supply tapping it could be raised to 56%. In practice, with a suitable transformer voltage setting and with the isolating switch closed, the start-run changeover switch is operated in both modes to give a full supply voltage to the motor's stator windings. Again, the motor will be provided with the necessary overload protection. Both this method and the previous method can be carried out automatically using timing relays (see Figure 5.35).

One other a.c. method commonly used is *rotor resistance*, which is employed for those three-phase induction motors with wound rotors, normally called *slip-ring motors*. The method is often used for motors up to 100 kW and Figure 5.36 is typical of both stator and rotor resistance connections. Briefly, when the motor is first switched on, the external resistance must be fully in; as the speed increases, its starting torque decreases and it will settle at a steady speed when its own torque and load torque are equal. As more and more resistance is cut out, the motor's torque will continue to increase and eventually it will settle down to its normal constant running speed. The motor starter will be provided with all the necessary interlocks to ensure that it cannot be switched on with the external resistance on the wrong tapping at starting; there will also be provision to short circuit the slip-rings in order to allow the motor to run as a normal cage induction motor. The advantages of this method of starting is that it provides a much higher

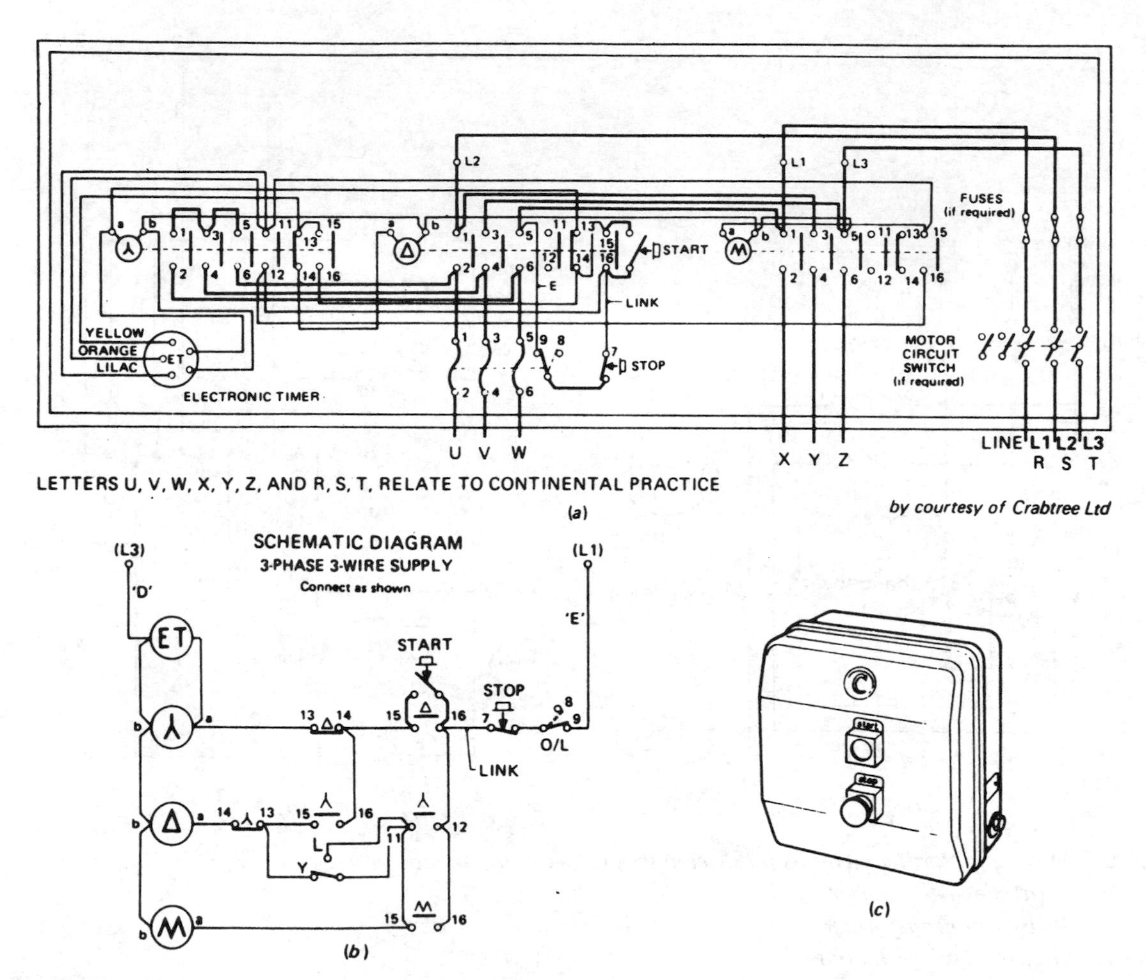

Figure 5.35 *Automatic star-delta starter: (a) circuitry; (b) control circuit; (c) physical appearance*

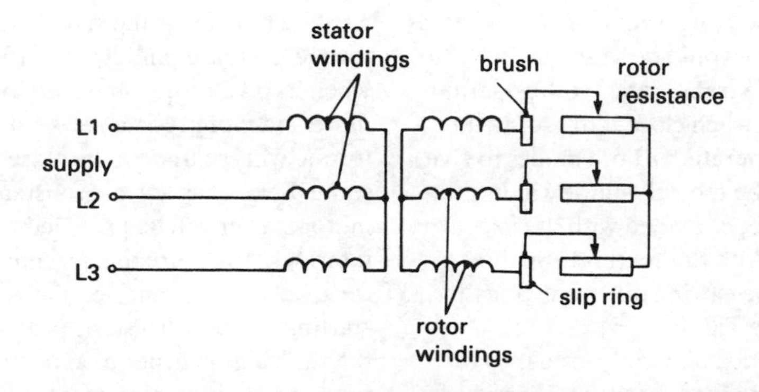

Figure 5.36 *Rotor resistance for a slip-ring wound rotor, induction motor*

starting torque and lower starting current than the above cage rotor induction motors. It also provides a degree of speed control by means of the external rotor resistance.

The comparison between a.c. and d.c. types of motor can possibly be best described by saying that, in either case, they both have *two* essential parts, namely a *fixed part* and a *rotating part*. In a.c. motors the fixed part is known as the *stator*, which is supported by a casing or outer frame (see Figure 5.37). The stator core consists of windings embedded in slotted laminations called stampings. The casing is often called the motor's *yoke* and this is also the name given to the frame in d.c. motors where, in this case, it supports the main *electromagnetic field poles*. Thus, the casing or yoke supports the motor's main field windings, which are connected to the supply. The rotating part in an a.c. motor is called the *rotor*, whereas in the d.c. motor it is called the *armature*. This is the main terminology difference between the two types of motor. A motor cannot work unless it has a second field system to interact with, and in the case of the a.c. induction motor this is created by converting the rotor into a temporary magnet by

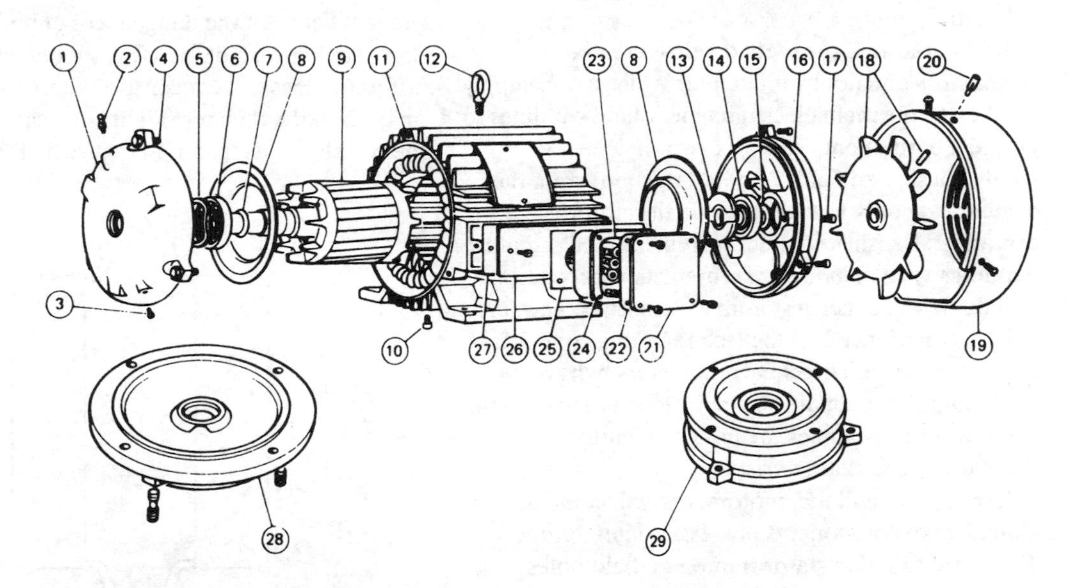

Figure 5.37 *Components of a typical cage-rotor induction motor*

1 Endshield, driving end
2 Grease nipple
3 Grease relief screw
4 End securing bolt, or through bolt and nuts
5 Anti-bump nuts
6 Ball bearing, driving end
7 False beating shoulder
8 Flume
9 Rotor on shaft
10 Drain plug
11 Yoke with or without feet
12 Eyebolt
13 Inside cap, non-driving end
14 Ball bearing, non-driving end
15 Circlip
16 Endshield, non-driving end
17 Inside cap screws
18 Fan with peg or key
19 Fan cover
20 Lubricator extension pipe
21 Terminal box cover
22 Terminal box cover gasket
23 Terminal board
24 Terminal box
25 Terminal box gasket
26 Raceway plate
27 Raceway plate gasket
28 D flange
29 C face flange

means of *electromagnetic induction*. The rotor is, in effect, a lump of metal called a *cage*; when the motor is switched on, its rotor acquires a north- and south-seeking pole by *induction* and starts chasing after a travelling magnetic field created by the stator windings. In the case of the supply being three-phase, the travelling field will be natural since the supply will already have red, yellow and blue phases automatically revolving at *synchronus speed* (n_s). The rotor, on becoming induced, tries to catch up with the rotating flux, but is unable to and eventually settles down to run at asynchronous speed or *rotor speed* (n_r). The difference between these two speeds is called the *slip*(s).

In d.c. motors, the second field is created by *conduction* and not by induction. Whilst a d.c. supply is taken to the main electromagnetic field windings to create a permanent cross-flux (which does not rotate), a second field is created by connecting the same d.c. supply to *brushgear* on the motor's armature. Brushes conduct current into the armature windings via a *commutator*. Interaction between the two fields will occur just as in the a.c. case, and force will be experienced on the free revolving conductors of the armature. Direct current motors behave according to their main field connections and only the shunt motor resembles the induction motor's constant-speed characteristics.

In both a.c. and d.c. motors, one will come across a number of components possessing *laminations*. These are found in stator stampings, field poles, armatures and rotors. Such laminations are a means of reducing *eddy currents*, which cause overheating; this is only one of a number of losses in a motor that lower its overall efficiency.

Motor fixing and maintenance procedures

It is essential to see that motors are correctly fixed in position. Foundation bolts and bedplates must be grouted in well by using a good mix of cement (usually two parts sand and one of cement in an easy running consistency). When foundation bolts have been set in position, checks should be made on the bedplate for parallel and diagonal level. It may be necessary to adjust the amount of packing beneath the bedplate. After this, the foundation bolts can be tightened down and preparation for grouting the bedplate can commence. It is preferable to erect a framework or shuttering around the base of the bedplate. The grouting mixture is then poured inside and left to set. The next step is to remove the shuttering and give a smooth finish to the grouting and foundation surfaces.

Where a motor requires to be directly coupled to a driven machine such as a compressor or pumpset, it is essential to ensure that both machine shafts are in line. One method is to insert callipers (or feeler gauges) between the flange faces of both shafts, as shown in Figure 5.38(a). This should be done at different places. If concentric alignment is required on both shafts then a straight edge can be placed on the couplings, as shown in Figure 5.38(b).

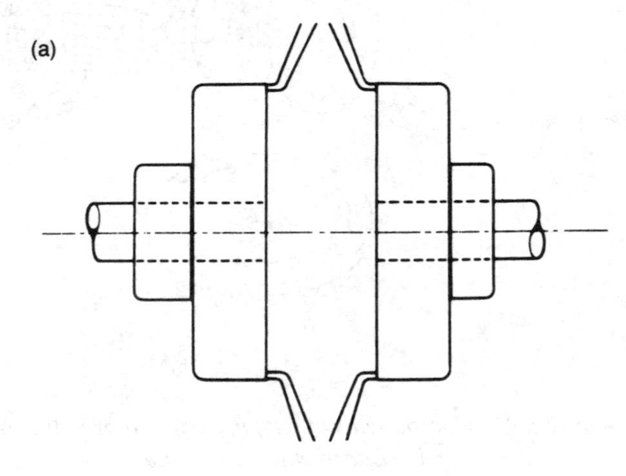

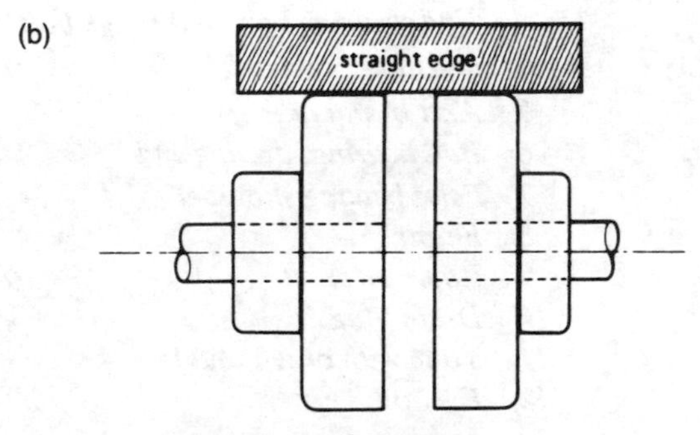

Figure 5.38 *(a) Shaft alignment*
(b) coupling alignment

Where a motor is to be coupled to a machine using belts, a slide rail should be used, as shown in Figure 5.39. The drive should be arranged with the slack side uppermost. A cord is run along the face of the large driven pulley and the motor is adjusted in position. A straight edge can be used to check if this is correct. Where vee belts are used, it is important to see that they are all of the same length. It is also important to check for belt stretch after the initial running.

With regard to maintenance, checks will need to be made on both motor and machine foundation bolts to see if they have worked loose during service. Care is needed when fitting pinions, pulleys and couplings, and they should not be physically driven on to a machine shaft. Misalignment can lead to vibration, bearing wear and eventually mechanical breakdown. Bearings should be checked for lubrication and for wear, which might cause the stator and rotor to rub together. Checks should be made for dust deposits around the motor windings and casing, particularly around the vent openings. A programme of maintenance is very important and motor manufacturers' data should be correctly kept and filed. It is not uncommon to find that careless servicing causes lost production. A few general hints on fault tracing are given below, in the order stated.

Trouble

1 Motor connected correctly but will not start.
2 Motor will not reach full speed.
3 Motor overheats on load.

Cause

1 Supply failure (either complete/one phase gone/overload).
2 Overload or voltage drop in cable.
3 Overloading or single phasing.

Remedy

1 Disconnect immediately and check supply to motor. If overloading, reduce load.
2 Reduce load or use larger motor. Check voltage and use larger cable.
3 Check motor temperature. If overloading, reduce load. Check supply.

Finally, maintenance procedure should also extend to control equipment such as starters, contactors and safety equipment, including guards.

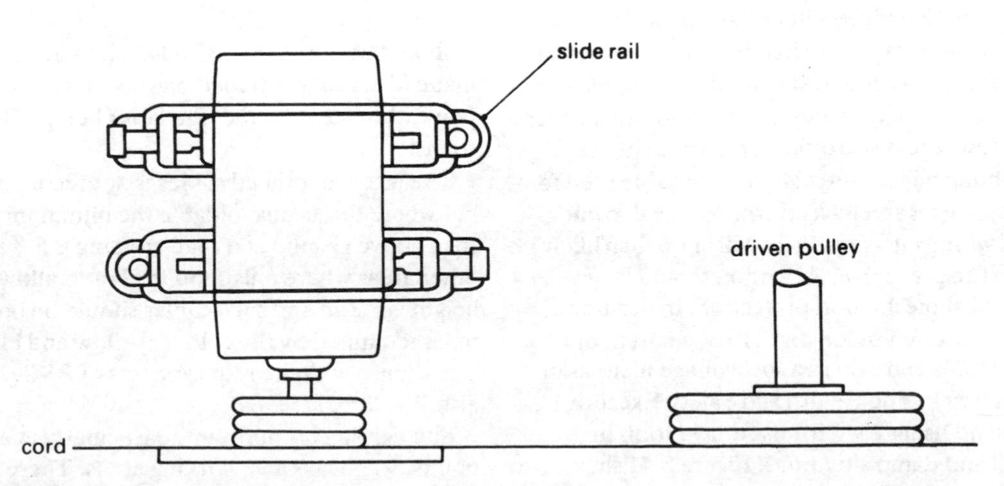

Figure 5.39 *Motor mounted on slide rail*

Special installations

Under this heading, the following installations will receive a mention:

(a) temporary/construction site installation,
(b) agricultural/horticultural installations,
(c) flammable/explosive installations.

Temporary/construction site installations

Construction sites are, by definition, temporary installations and although site work was mentioned in the chapter dealing with industrial studies, here the subject is taken a little further. In practice, a large project will require an electrical supply as soon as possible and site distribution needs should be planned well in advance. It is possible to find a generator being used where no site supply is available; special sound-proof sets are designed to reduce excessive noise. Site distribution units should be designed to BS 4363 and such units should be placed in ready access for operatives so that the supply can be switched off in the event of an emergency. It should be noted that BS 4343 plugs and sockets are interlocked and should not be used for emergency switching. Figure 5.40 shows a typical arrangement of supplying BS 4363 distribution units from a single-phase supply. If a three-phase supply was required then the transformer in TU/1 would become 415 V/110 V and the unit would be renamed TU1/3.

It is likely that there will be fixed plant and moveable plant as well as lighting and small power needs for site offices and site buildings. Supplies to cranes, hoists and compressors may require a voltage of 415 V, whereas site offices and some site floodlighting will require 240 V. The cable used for these supplies is steel-wired armoured cable but internal wiring into site offices will more than likely be PVC/PVC/c.p.c. cable. Attention should be given to additional mechanical protection. In hazardous areas it is recommended that 110 V be used for portable tools and an extra-low voltage in areas of additional risk. The Health and Safety Executive recommend using 25 V for hand-held tools in confined and damp situations. Figure 5.41 shows two methods of providing additional safety: (a) using Class II equipment requiring no earth; and (b) using a system known as earth monitoring, where any break in the trailing lead will cause loss of supply to the circuit. Figure 5.42 shows a typical portable transformer circuit.

British Standards CP 1017 recommend the use of double-pole switching for controlling single-phase supplies up to 110 V; there should also be circuit protection provided in each live conductor.

Socket outlets and plugs should be designed to BS 4343, have a rating not exceeding 16 A and be mechanically interlocked to ensure that the supply to the contact tubes of the accessory is isolated when the plug is withdrawn (see Figure 5.43). Accessories are available that provide discrimination between different voltages by colour coding and by the positioning of the earth contact in relation to the keyway (see Figure 5.44). The colour coding is given in the following table.

Voltage	*Colour*
750 – 500 V	black
415 – 380 V	red
240 – 220 V	blue
130 – 110 V	yellow
50 V	white
25 V	violet

It should be noted that BS 4343 plugs and sockets are made of either a moulded plastic material or a polycarbonate material, the latter being extremely robust.

The use of overhead cables is not recommended, but where this is unavoidable the minimum height of span above ground at a road crossing is 5.8 m. In other areas where only mobile plant is allowed, the height is 5.2 m and such cables should be bound with tapes identified by the colours yellow and black in accordance with the requirements of BS 2929 (see also BS CP 1017: 1969).

Site lighting is important, particularly at access points, walkways and working areas. There should also be warning lights fitted to obstructions to remind

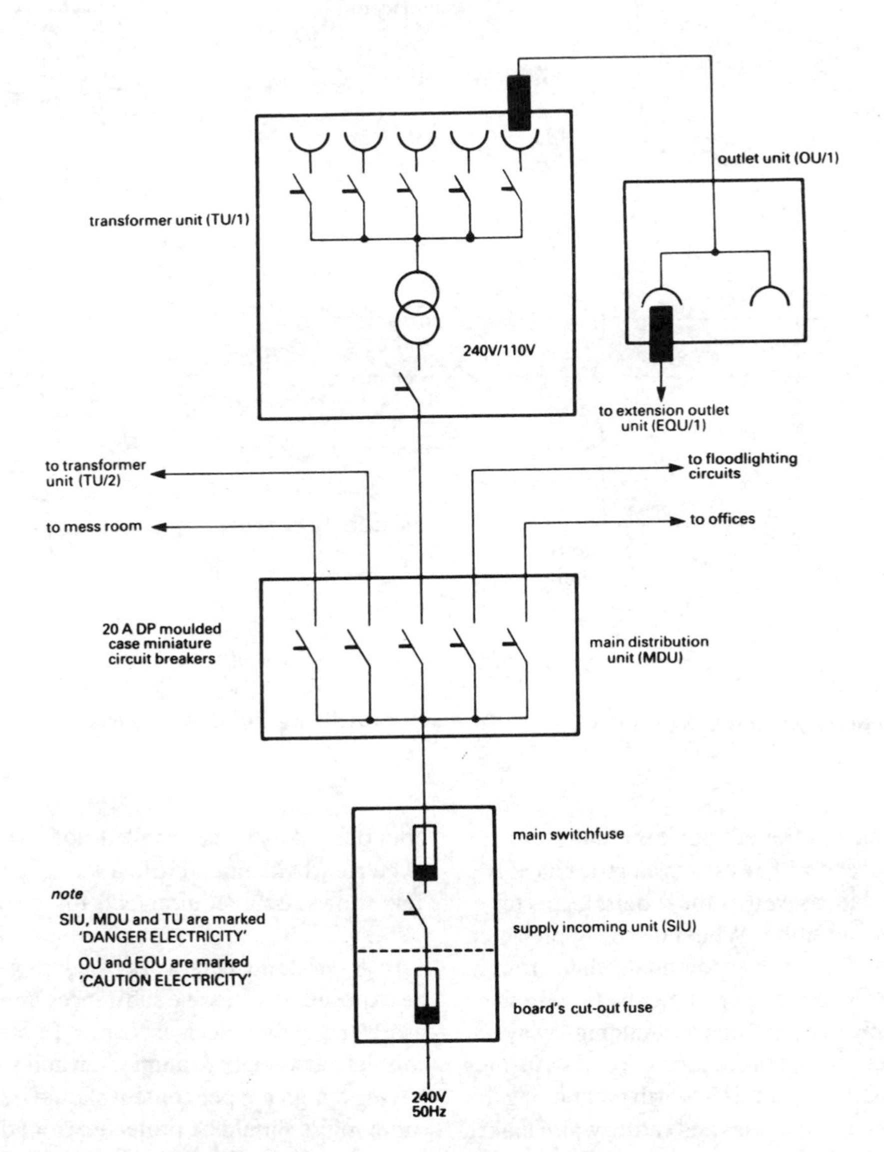

Figure 5.40 *Construction site supply*

personnel of the hazard. Security lighting is also recommended. Temporary installations should be inspected and tested at intervals not exceeding three months.

Agricultural/horticultural installations

There are several requirements for agricultural and horticultural installations given in the IEE Regulations, namely Section 605. Class II equipment

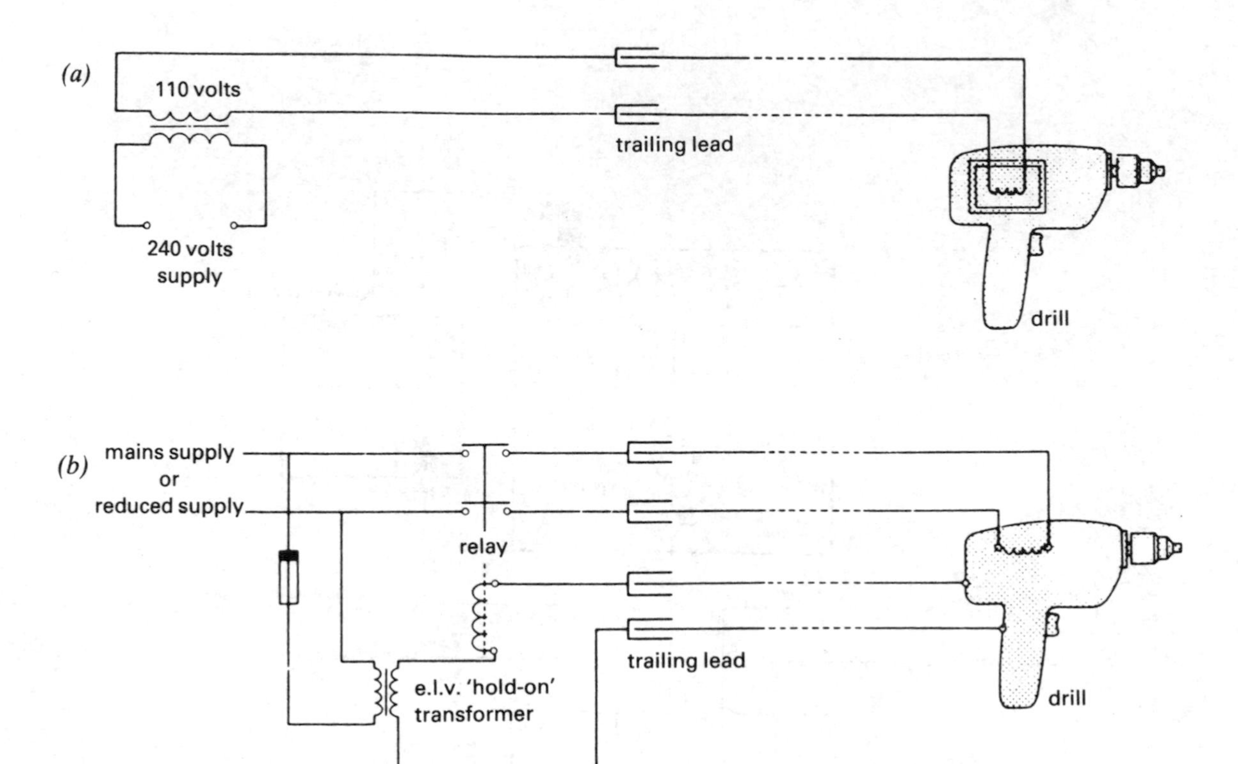

Figure 5.41 *(a) Supply to a double insulated drill (b) Earth monitoring supply to an electric drill*

has already been mentioned, but it is recommended that BS 2632 be read with regard to electric fences and the types available, as well as the requirements for their operation and siting. What must be appreciated is that these installations are potentially dangerous in damp, wet and corrosive environments. Provision for an electric supply in numerous outbuildings away from the original zone of protection becomes a further problem and risk situation. It is worth mentioning the low body resistance of horses and cattle, which makes them susceptible to electric shock at voltages lower than 25 V a.c. British Standard 2632 calls for electric fence controllers to be properly designed, with no possible contact between input and output terminals. Controllers today are often battery operated, with weatherproof and double-insulation casings. One end of the controller's output is connected directly to an earth electrode while the other end is connected to the fence. Low-energy pulses are transmitted at 1.5 ms to deter any curious animals trying to go beyond the boundary. Any fence installed along a public road has to be clearly identified with a warning sign. Figure 5.45 shows some requirements for a typical farm installation.

In general, no metalwork of a wiring system should be exposed to corrosive substances unless it is protected against such exposure. In damp situations, contact between bare aluminium and any other metal having a high copper content should be avoided. All cable joints should be protected from the ingress of moisture, and cable buried in the ground should preferably be armoured, have a metal sheath or be of the PVC-insulated concentric type. They should be buried at a safe and sufficient depth to avoid damage, and their presence should be identified by marker tape or cable covers. Where an 'all-insulated' wiring system cannot be used, consideration should be given to the use of a reduced voltage for portable tools. Residual-current devices should be encouraged where danger exists.

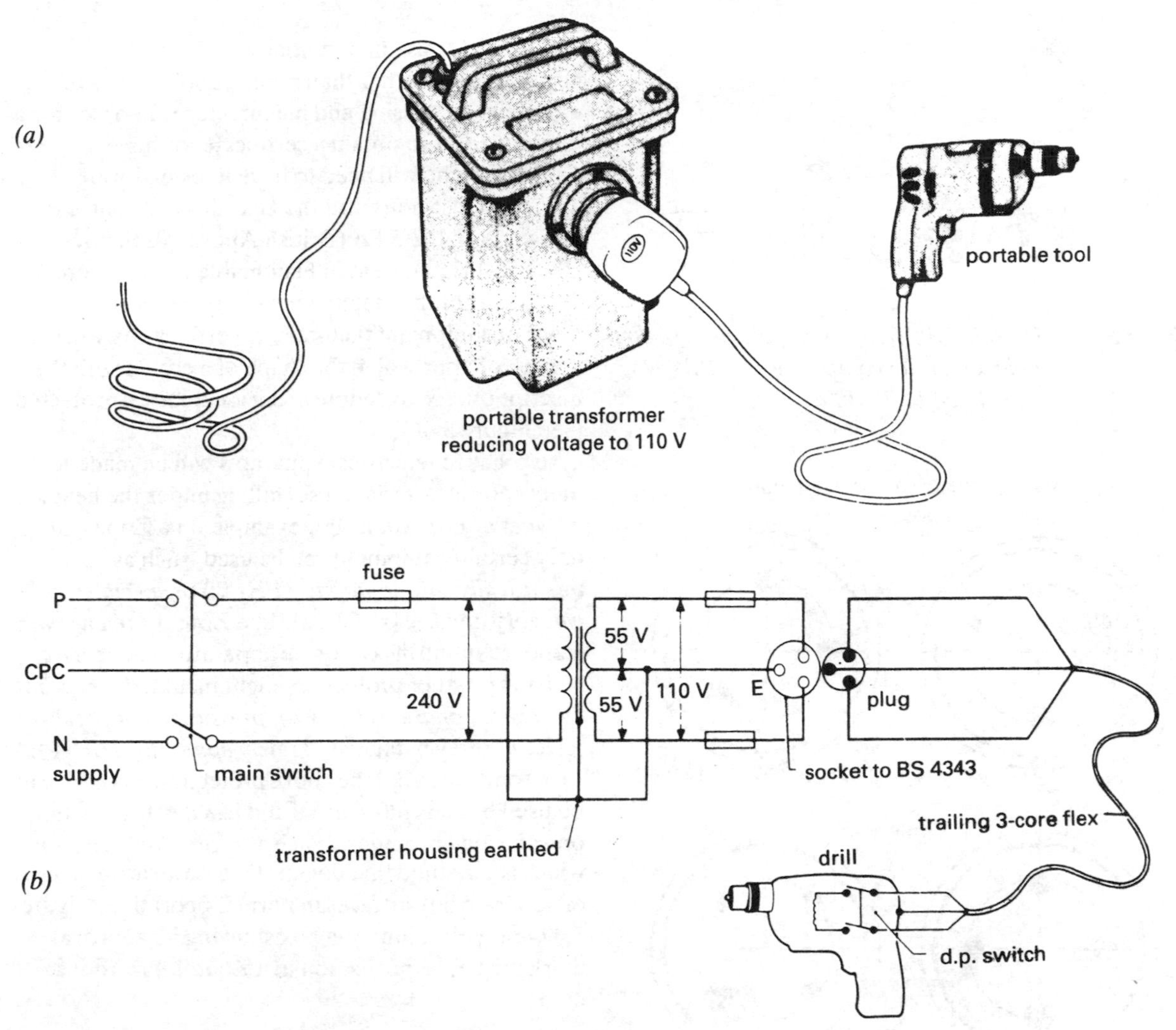

Figure 5.42 *(a) Method of reducing voltage on site (b) Wiring to a portable drill through 240/110 V step-down transformer*

Flammable/explosive installations

Flammable and explosive installations will be required in premises that have been designated hazardous areas, such as petrol-filling stations, chemical works, grain stores and even offshore oil rigs. Hazardous areas are classified into zones of risk, in order of severity:

Zone 0—in which an explosive gas–air mixture is continuously present, or present for long periods

Zone 1—in which an explosive gas–air mixture is likely to occur in normal operation

Zone 2—in which an explosive gas–air mixture is not likely to occur in normal operation, and if it does occur, it will exist only for a short time.

There is an exhaustive list of flammable material that can easily burn if ignited or possibly explode if it is mixed with air. The lowest temperature at which sufficient heat is given off from a flammable substance resulting in an explosive gas–air mixture is

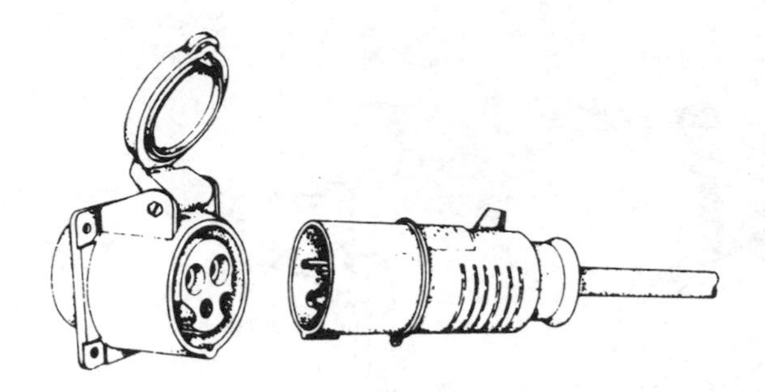

Figure 5.43 *Reyolle 240 V/16 A two-pole earth socket outlet and plug complying with BS 4343, CEE 17 and IEC 309*

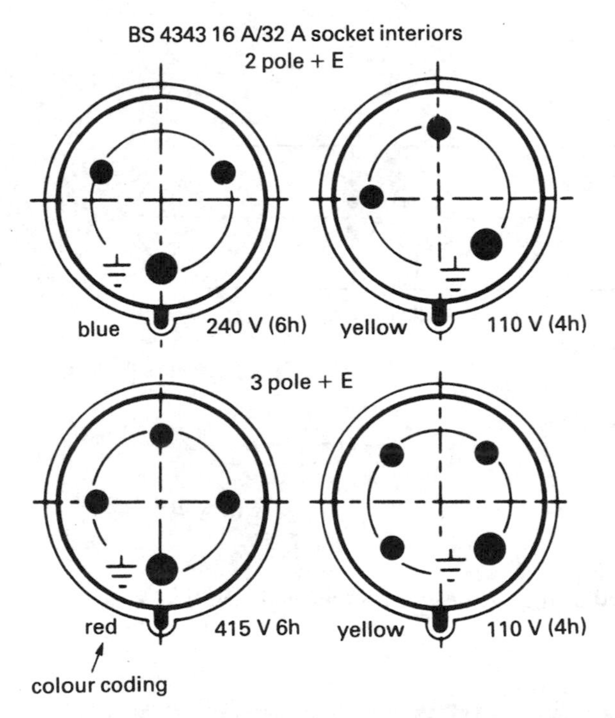

Figure 5.44 *Voltage discrimination for BS 4343 sockets*

called the *flashpoint*. Moreover, explosive mixtures differ considerably, e.g. a mixture of air and town gas will ignite at 560°C whereas a mixture of air and petrol will ignite at 250°C.

To avoid such risk of explosion, electrical equipment has to be designed so that it is capable of being used in such areas. Most manufacturers of equipment for this purpose will be involved with BS 5345, which is the code of practice for the selection, installation and maintenance of electrical apparatus for use in potentially explosive atmospheres—other than mining applications or explosive processing and manufacture. In order for a manufacturer to obtain a certificate for his equipment, he will need to have it tested by an approved test-house. In the UK, this test centre is known as *BASEEFA* (British Approvals Service for Electrical Equipment in Flammable Atmospheres) and is part of the Health and Safety Executive. A piece of equipment that satisfies certification will bear a mark of approval in the shape of a crown with the inscription '*Ex*' to denote that it is explosion protected (see Figure 5.46).

In practice, electrical apparatus will be made to meet certain criteria of use, falling under the heading of *types of protection*. For example, in a Zone 0 area only certain equipment can be used, such as *intrinsically safe apparatus (EEx ia)* or *special category apparatus (EEx s)*. In a Zone 1 area a lower standard of intrinsically safe apparatus can be used *(EEx ib)*. Other protection might include *flameproof enclosure apparatus (EEx d), pressurized apparatus (EEx p)* or even *increased safety apparatus (EEx e)*. In a Zone 2 area, all the above protection methods can be used but it is more usual and less costly to install one of a lower standard, such as *protection n (EEx n)*, which is a method that does not create an arc or spark, or generate hot surfaces in normal operation. Figure 5.47 is a typical junction box showing its approval mark for EEx e protection in a Zone 1 hazardous area.

Exercise 5

1 Figure 5.48 shows a circuit diagram of a 30 A ring final circuit consisting of six 13 A BS 1363 socket outlets. Redraw the diagram and, where possible, incorporate further socket outlets at: (i) the origin of the circuit; (ii) a socket outlet on the ring; and (iii) a joint box half-way around the ring.

2 Describe the internal components of a metal-clad six-way consumer unit fitted with miniature circuit breakers.

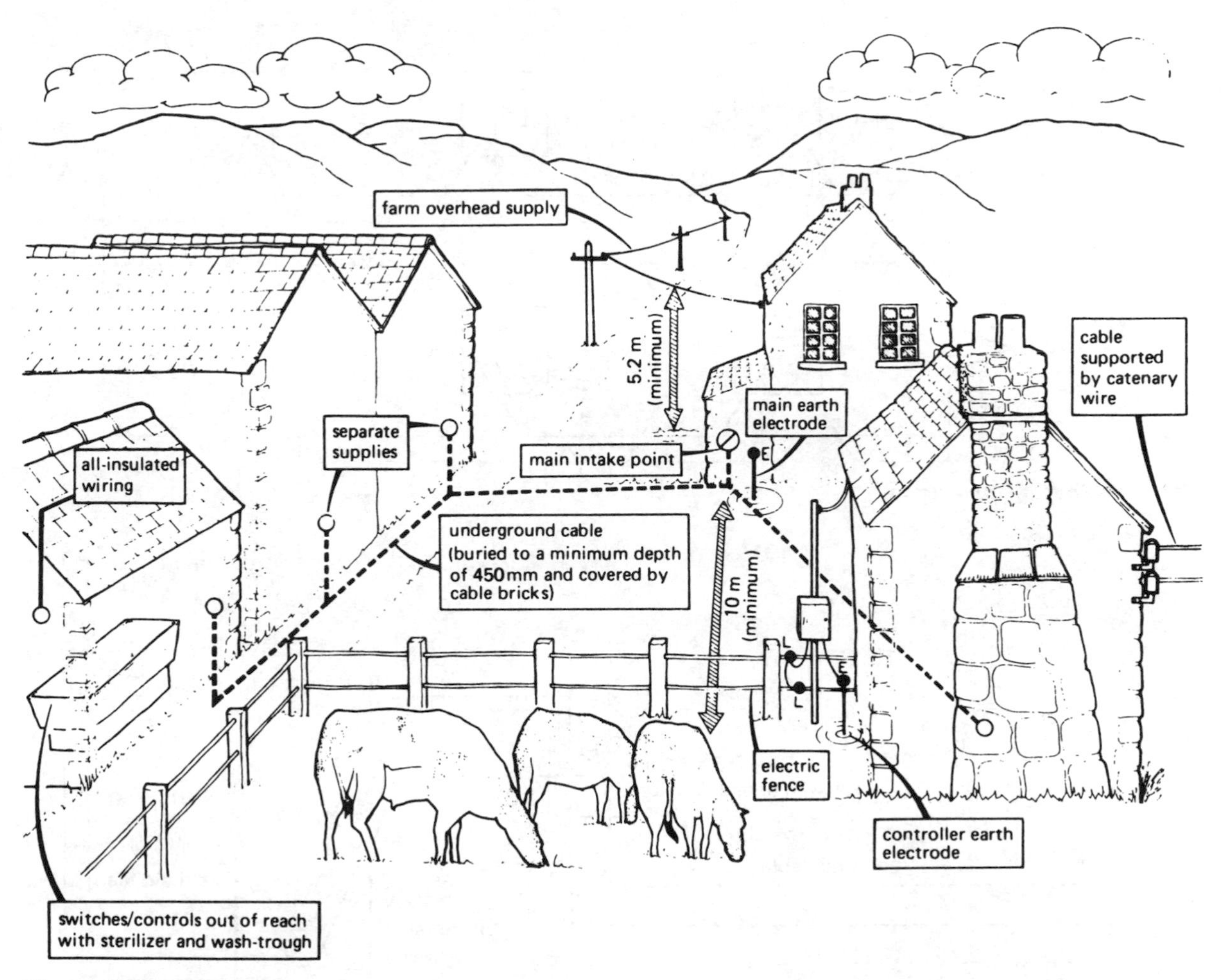

Figure 5.45 *Farm wiring*

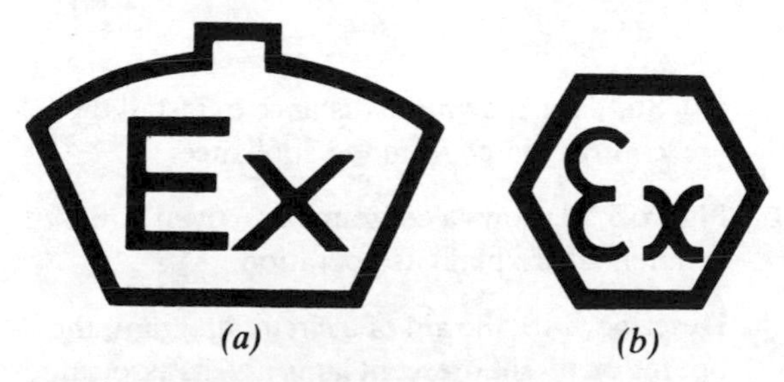

(a) *(b)*

Figure 5.46 *expolosive equipment markings*
(a) BASEEFA mark
(b) EEC mark

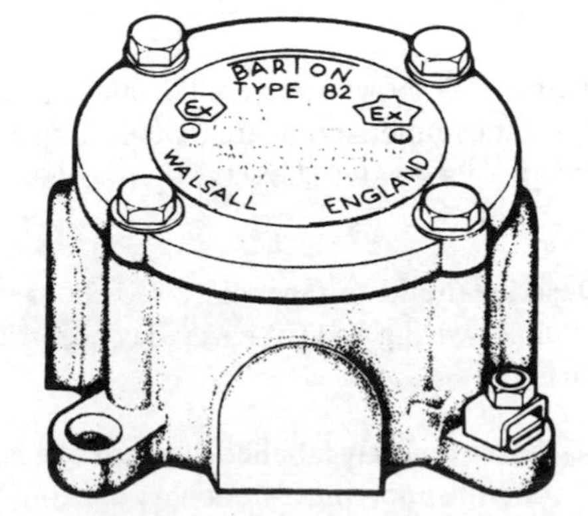

Figure 5.47 *'Barton' weatherpoof junction box providing increased safety (Ex 'e') protection*

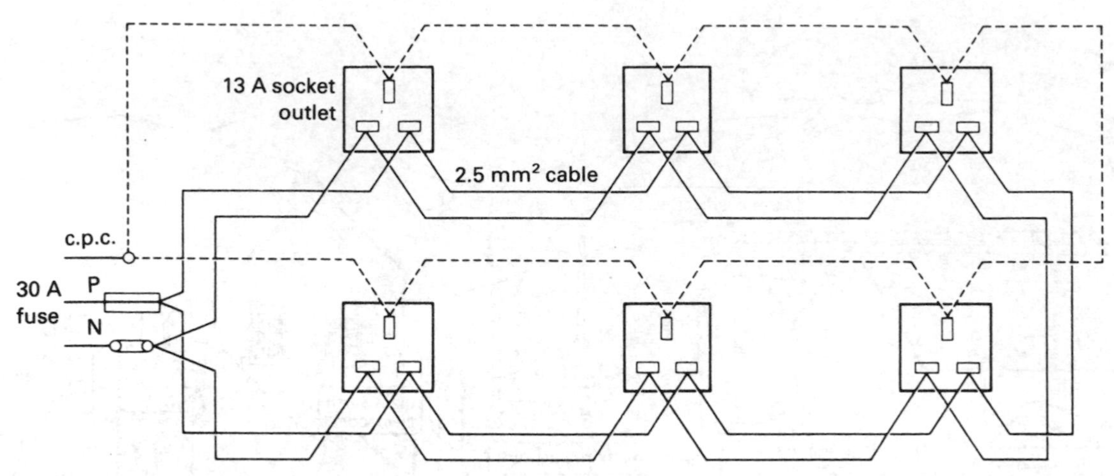

Figure 5.48 *Ring final circuit*

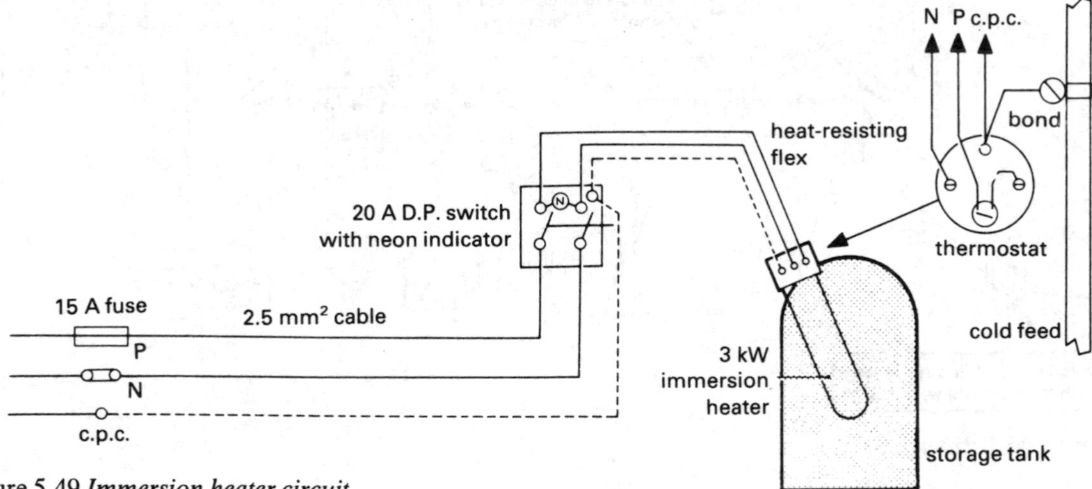

Figure 5.49 *Immersion heater circuit*

3 Figure 5.49 shows the circuit diagram of a domestic immersion heater circuit. Explain with the aid of a diagram how the thermostat operates.

4 Describe the advantages and disadvantages of (a) catenary wiring and (b) earthed concentric wiring.

5 (a) Draw a neatly labelled diagram of a final circuit supplying a stationary cooking appliance; include the internal connections of the control unit, which has facility for a 13 A socket outlet.
(b) State the maximum distance to install the control switch from the appliance.

6 Figure 5.50 shows a diagram of an oven hotplate simmerstat. Explain its operation.

7 Describe, with the aid of a circuit diagram, the operation of a fluorescent lamp and its associated control gear.

8 (a) What are the IEE Wiring Regulations covering motor circuits?
(b) Draw a circuit diagram of a single-phase a.c. motor controlled by means of a direct on-line contactor starter having thermal overloads.

9 (a) List five requirements of the IEE Wiring Regulations concerning rising main supplies.
(b) State *two* advantages of this wiring system.

10 Draw a neatly labelled diagram showing the means of control for a small factory having lighting, heating and power circuits fed from separate distribution boards.

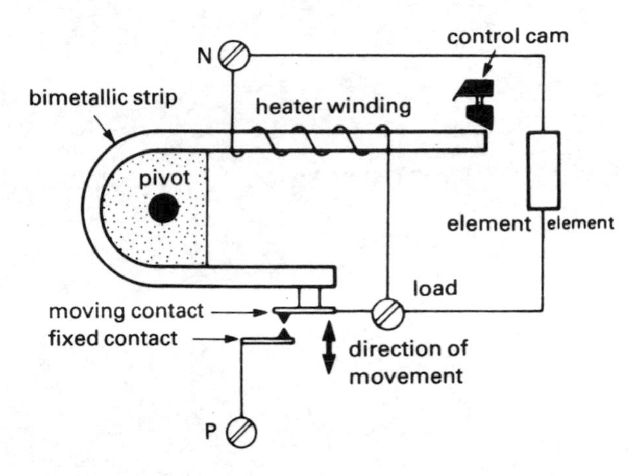

Figure 5.50 *Oven simmerstat*

chapter six

Inspection and testing

After reading this chapter you should be able to:

1 know the basic requirements for inspecting and testing consumers' premises,
2 state numerous inspection checks needed on electrical installations, required by the IEE Wiring Regulations,
3 describe several basic testing procedures required by the IEE Wiring Regulations, stating acceptable test results,
4 draw diagrams of several testing circuits showing the correct types of instruments used,
5 state the purpose and use of an ohmmeter and tong tester.

The main purpose of inspecting and testing electrical installations is to ensure that they are safe to use and also that they will function efficiently under both normal and abnormal conditions. It is to this end that the IEE Wiring Regulations require both visual inspection and testing to be carried out when the installation is finally completed. This has to be done in accordance with Reg. 130–10–01, which partly states:

On completion of an installation an extension or alteration of an installation, appropriate tests and inspection shall be made, to verify, so far as is reasonably practicable, that the requirements of Regulations 130–01 to 130–09 have been met.

Reference will need to be made to Part 7, Chapter 71 of the IEE Regulations. Where wiring alterations take place and periodic inspections and tests are made, they will need to comply with Chapters 72 and 73, respectively.

The regulation requires information about the installation; this has to be made available to the person carrying out the visual inspection and testing work. It should include diagrams, charts, tables and details of circuits, including relevant design data. The designer and the installer are both responsible parties for this final stage before the installation is commissioned and handed over to the client.

Inspection

Reg. 712–01–03 of the IEE Regulations provides a checklist for visual inspection of the installation and, whilst it may not apply in every circumstance, it must not be forgotten that the purpose is to see if compliance with the IEE Regulations has been met, particularly the requirements of Part 5—that equipment has been selected to British Standards, a foreign recognised standard or some specification endorsed by CENELEC. One is also looking to see that no installed equipment is damaged in any way as to impair safety.

For the majority of small installations, including domestic dwellings, and in the absence of electrical equipment provided by the client, the checklist can be reduced to:

(a) conductor checks on selection, connection and identification,
(b) presence of appropriate devices for isolation and switching, including circuit protective devices and earth leakage protective devices,
(c) presence of labels, diagrams, charts, warning notices and notice of instruction.

In (a) above, the areas will include checking to see if cables and conductors are of the correct size, that they are correctly labelled and identified, and that they are connected to their correct terminals at lampholders, switches and socket outlets. It is important to check that the connection of single-pole switches are in the phase conductor. There should be no undue stress imposed on cable/conductor connections. Checks should be made to see if steps have been taken against possible damage likely to be caused by heat, dampness and/or corrosion, and also to see that insulation has not been removed further than necessary.

In (b), it has already been pointed out in the chapters on installation (Chapters 4 and 5) that every installation needs to have a means of isolation and a means of excess current protection, as well as earth leakage protection. Most switchgear will incorporate a means of control and protection, but it is also important to check if final circuits are also protected and switched independently. Access to switching and control devices is very important. Some socket outlet circuits are fitted with residual current devices, the presence of which should be noted.

In (c), checks should be made to see if labels and charts are provided in distribution boards and consumer units, as well as in other places where information might be required, e.g. in other buildings. Reference should be made to Section 514 of the IEE Regulations regarding notices. A notice of periodic inspection and testing is required by Reg. 514–5 and has to be fixed in a prominent position at or near the main distribution board. This also applies to installations controlled by residual current devices (see Reg. 514–12–02).

In the larger installations, checks will be made for the presence of fire barriers and protection against thermal effects, as well as those pertaining to methods of protection used against the possiblity of direct contact (including measurements of distances, where appropriate). A number of the checks made are normally covered in the design specification or shown on the drawings, e.g. selection of equipment and protective measures appropriate to external influences, and the choice and setting of protective and monitoring devices.

Testing

This is carried out to ensure that conductors have been installed and connected correctly. The IEE Regulations provide a list of the tests to be made, which are required to be carried out in the sequence indicated (Section 713). For smaller installations, the main tests are:

(a) continuity of protective conductors, including bonding, ✓
(b) continuity of ring final circuit conductors,
(c) insulation resistance, ✓
(d) polarity, ✓
(e) earth fault loop impedance,
(f) earth electrode resistance,
(g) operation of residual current devices.

Testing requires having those instruments available that are capable of giving reliable readings. Such instruments should be used carefully and be protected on site, as well as periodically tested for accuracy.

Continuity of conductors implies using an instrument that can read low values of resistance (between 1 and 0.005Ω). A 30 m length of 2.5 mm²/1.5 mm² PVC/PVC/c.p.c. cable will only have a resistance of 0.33Ω for the phase conductor and 0.55Ω for the c.p.c. conductor. It is preferable to have a multirange instrument with two scales, as recommended by the ECA Technical Committee, e.g. one that can read values between 0.001 and 2Ω and between 0.01 and 20Ω. Several manufacturers now make testers for testing protective conductors made of steel enclosures; the tests can be made with either a.c. (at mains frequency) or d.c., at a voltage not exceeding 50 V and with a current approaching 1.5 times the circuit design current, up to a maximum of 25 A.

There are two methods for carrying out *ring circuit continuity* tests. Method 1 uses a wandering lead of known resistance, whereas the second method does not and makes all the necessary tests at the main distribution board, using actual circuit conductors. Figure 6.1 shows the ohmmeter connections to both phase conductors. The first test (*Test A*) is to measure the resistance and then to repeat the test for both neutrals and both c.p.c. conductors. If the wiring system is in conduit then all the ring circuit conductors may be of the same size and therefore all the readings will be the same. However, if PVC/PVC/c.p.c. cable is used then the c.p.c. will have a higher reading, since it is a smaller conductor.

The second test (*Test B*) is to reconnect the phase conductors in the distribution board and take the wandering lead over to the nearest mid-point socket outlet on the ring (see dotted line). The resistance is again measured and the test repeated for the other two circuit conductors. Since the resistance of the wandering lead is known, it should be subtracted from

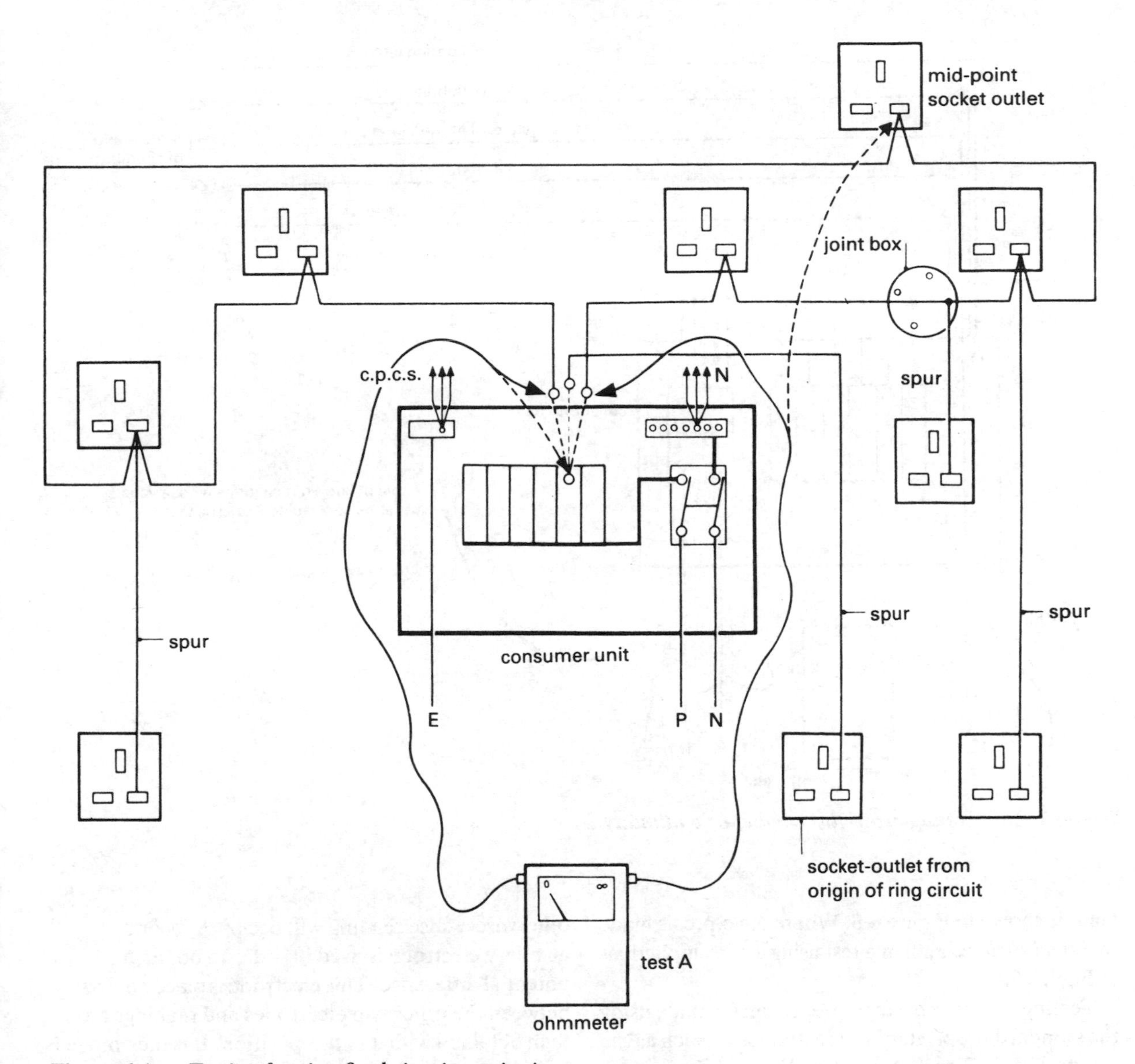

Figure 6.1 *Testing for ring final circuit continuity*

the second test results. This test method is satisfactory if it can be verified, for each circuit conductor, that *Test B* is about one-quarter of the reading of *Test A*.

In Method 2, *Test A* above is repeated and the conductors are reconnected back together. *Test B* involves shorting out the L.N.E. terminals of the mid-point socket outlet and making *two* further tests from the distribution board. The first is between phase and neutral and the second is between phase and c.p.c. This test method is satisfactory if the phase and neutral test is approximately one-half the value obtained for the *Test A* result involving the phase or neutral conductors. For the phase to c.p.c. test, the resistance should be approximately one-quarter of the *Test A* phase or neutral result *plus* one quarter of the resistance value of the *Test A* c.p.c. result.

The test for *continuity of copper protective conductors* is shown in Figure 6.2 using a d.c. ohmmeter and a cable of known resistance which again has to be subtracted from the initial readings. An alternative method to this, without a wandering

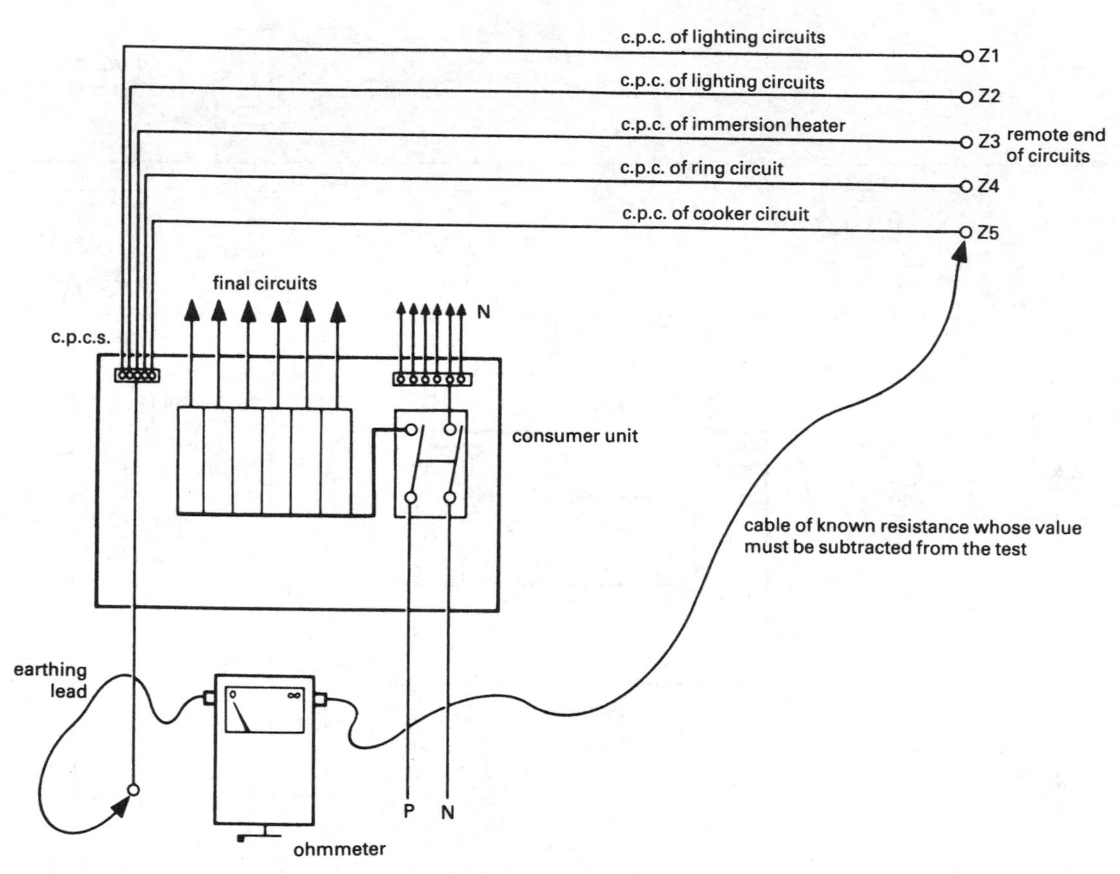

Figure 6.2 *Testing protective conductor continuity*

lead, is shown in Figure 6.3. Where the c.p.c. is made of a *steel enclosure*, then a test using a.c. is applied, as in Figure 6.4.

Testing *earth electrode resistance* can be made using the standard proprietary test instruments, such as the 'Megger' ET3/2 earth tester (see Figure 6.5).

Instruments are either battery-operated, hand-driven or mains-operated, and they are capable of reading low and high values of resistance.

Figure 6.6 shows how a test is made to measure the resistance of a consumer's earth electrode. It is made by passing a steady value on alternating current through the electrode (X) under test. Current will find its way towards the auxiliary electrode (Y) via the general mass of earth and return to the supply. The position of this latter electrode must be such that its resistance area does not overlap the test electrode, otherwise a false reading will occur. A second auxiliary electrode is used in order to obtain a potential difference. This electrode is placed half-way between the other two electrodes and readings are then obtained with it in this position. It is then moved 6 m either side of this position and further readings are taken. The average of these three readings will indicate the resistance value of the test electrode. It should be noted that for this test the main earth electrode must be disconnected from all sources of supply other than the test supply.

Insulation resistance tests are made with d.c. at least twice the nominal a.c. supply voltage, and it need not exceed 500 V for installations rated up to 500 V (i.e. 240 V/415 V a.c. premises). For installations above 500 V, then a 1000 V tester is required. This tester is ideal where a higher degree of safety is envisaged, such as with transient high voltages set up in motor

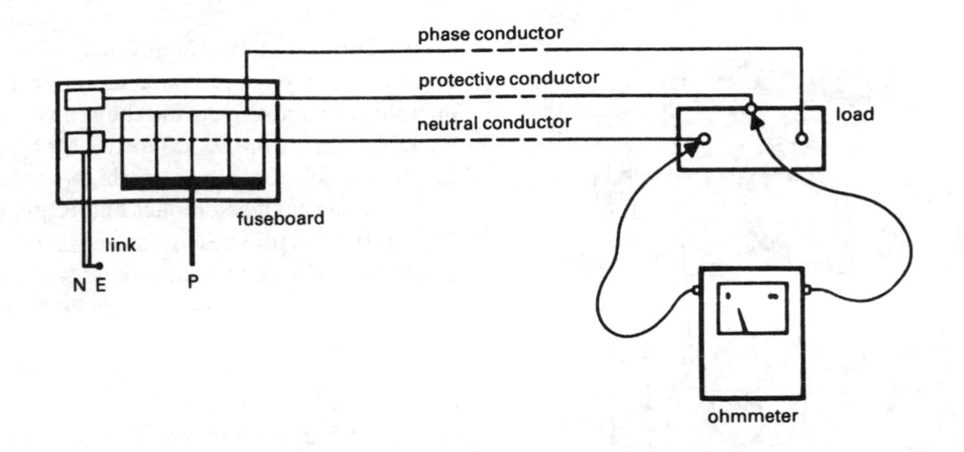

$$\text{resistance of protective conductor} = \frac{\text{neutral conductor c.s.a.}}{\text{protective conductor c.s.a.} + \text{neutral conductor c.s.a.}}$$

Figure 6.3 *Testing a protective conductor without using wandering test lead*

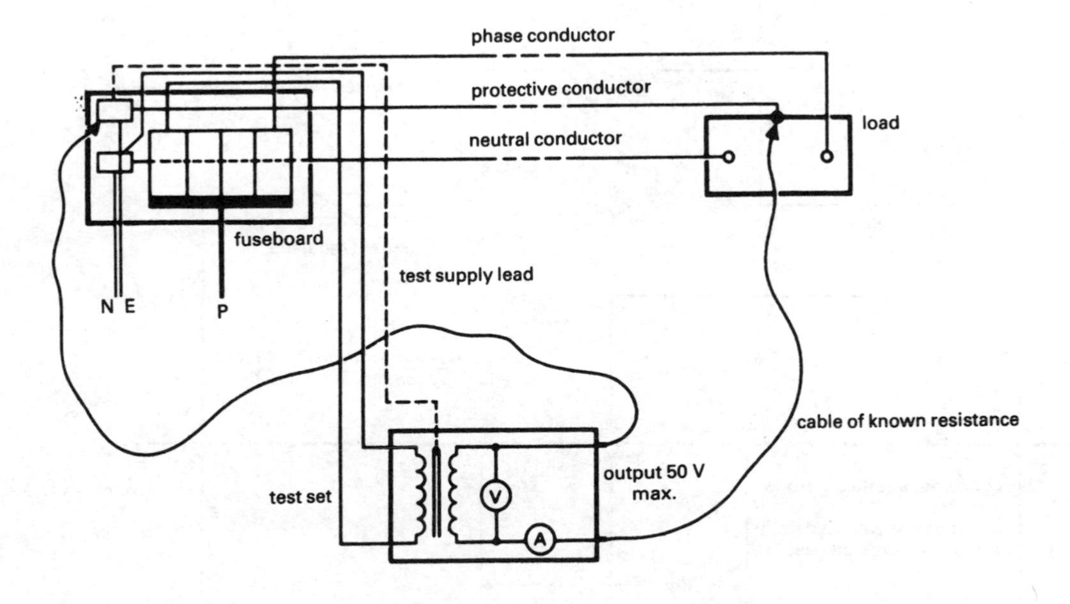

$$\text{resistance of protective conductor} = \frac{V}{I} \text{ ohms}$$

Figure 6.4 *Testing a protective conductor made of steel using alternating current*

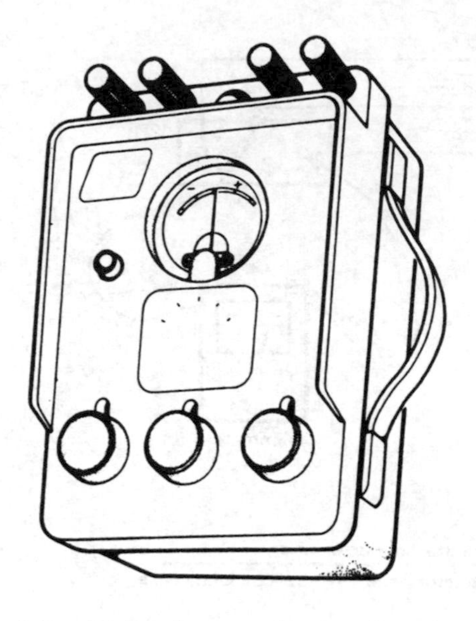

Figure 6.5 *Null-balance earth tester (battery operated)*

circuits and often found in m.i.m.s. cables. Dual-voltage instruments are available; they can be battery-operated or hand-driven and they can be analogue or digital, reading from 50 kΩ to 100 MΩ. Regulation 713–05–01 calls for the use of a high-voltage 'flash tester' on site-built assemblies and Regulation 713–08–01 calls for a test of insulation on non-conducting floors and walls—these will not be discussed. Electrical separation of circuits can be tested using a standard 500 V insulation tester. At least 5 MΩ is required of the insulation separating the circuits (See Reg. 713–06).

Figure 6.7 shows how *insulation resistance* tests are made on a completed installation. *Test 1* is made between both live conductors shorted out and the main circuit protective conductor. With all the circuit protective devices inserted and the main switch closed (assuming no supply), as well as local switches closed, the reading of the instrument should be not less than 0.5 MΩ for 240 V/415 V supplies. In *Test 2* the same result is expected, but this time all lamps and other

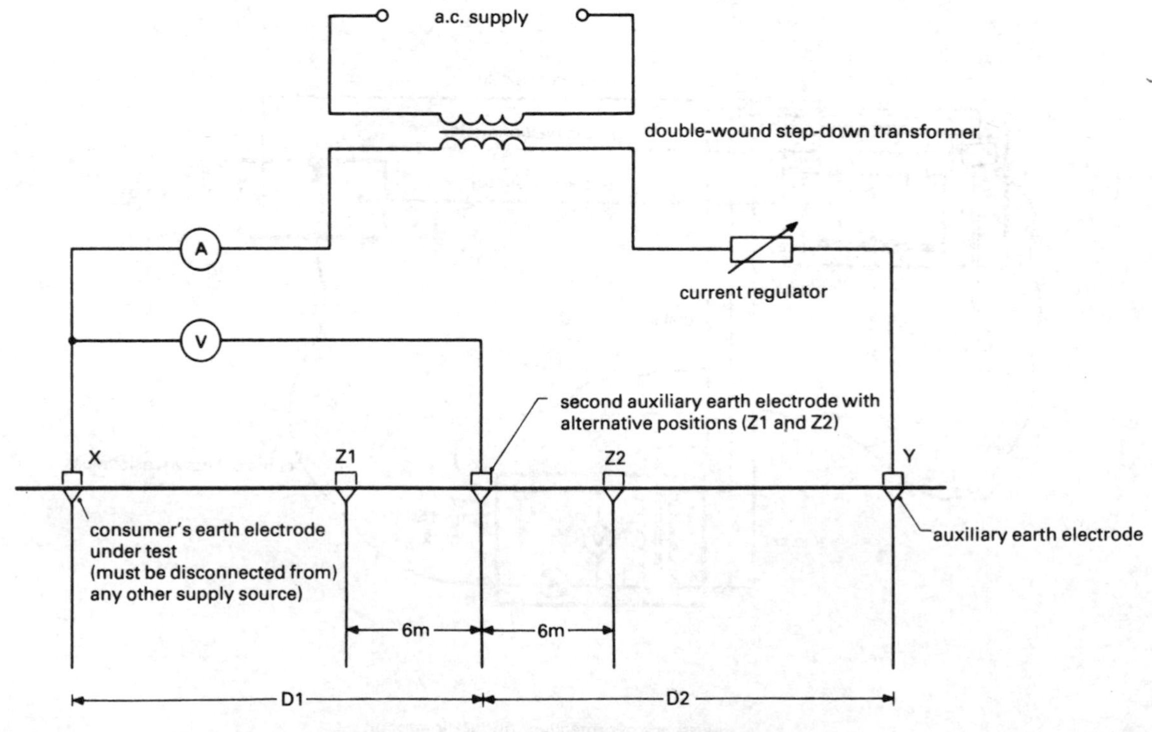

Figure 6.6 *Testing earth electrode resistance*

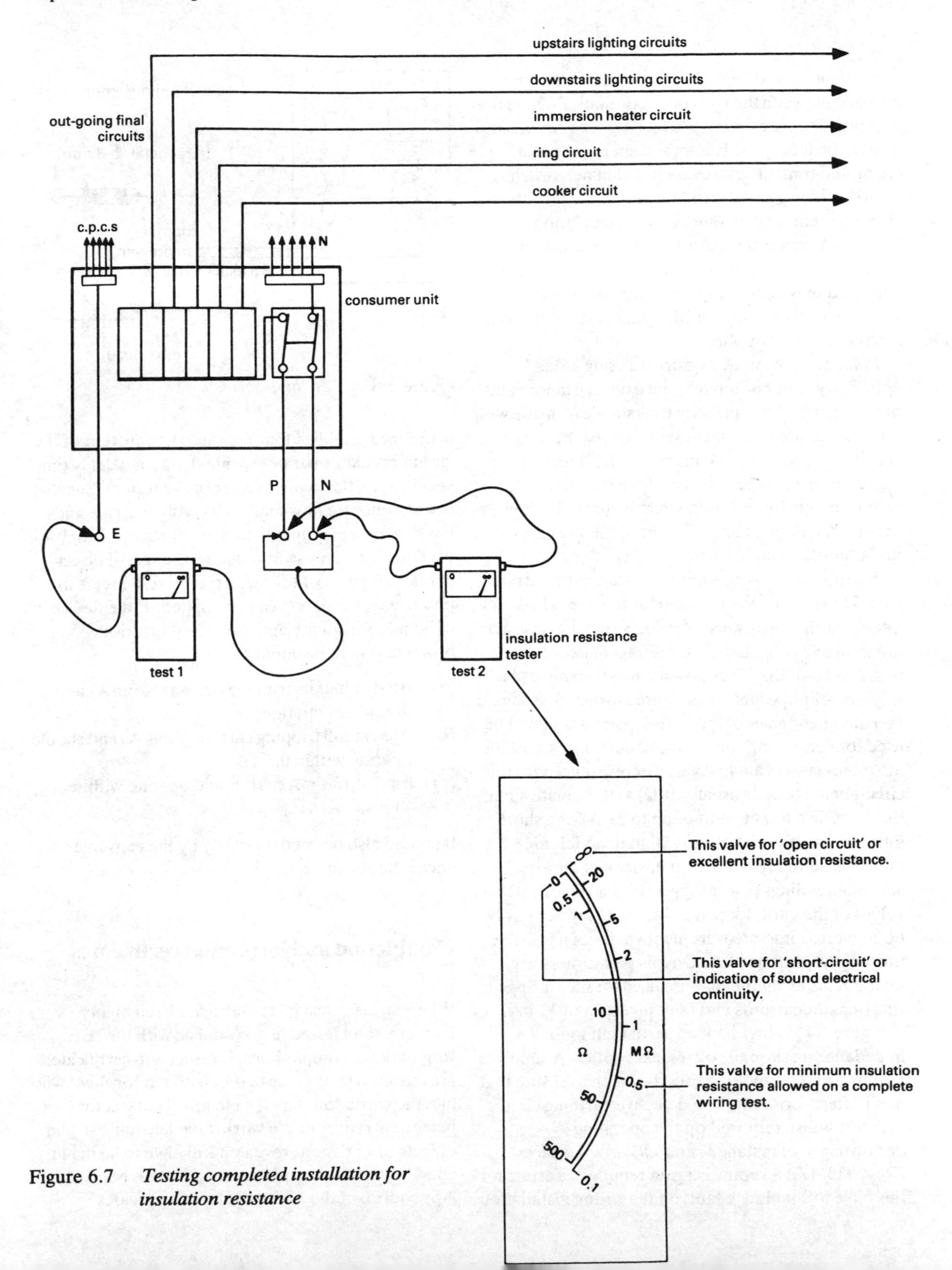

Figure 6.7 *Testing completed installation for insulation resistance*

equipment should be disconnected since the test will be made between the live conductors only. Where this is impractical, local switches feeding such circuits will need to be left open. It is important to note that electronic control devices such as dimmer switches could be damaged with a 500 V tester and they should either be removed or shorted out. Regulation 713–04–06 requires separate tests to be made of disconnected equipment having exposed conductive parts and they should obtain insulation resistance values of not less than 0.5 MΩ, unless an appropriate British Standard applies.

Polarity tests can be performed using a standard continuity tester or buzzer, since no resistance values are required. 'Live' polarity tests should be made with an appropriate instrument (see Figure 6.8). It is not recommended to use neon indicators, since they deteriorate with age. Figure 6.9 shows some of the requirements for polarity, while Figure 6.10 shows a method of finding continuity, insulation resistance and polarity of an MI cable.

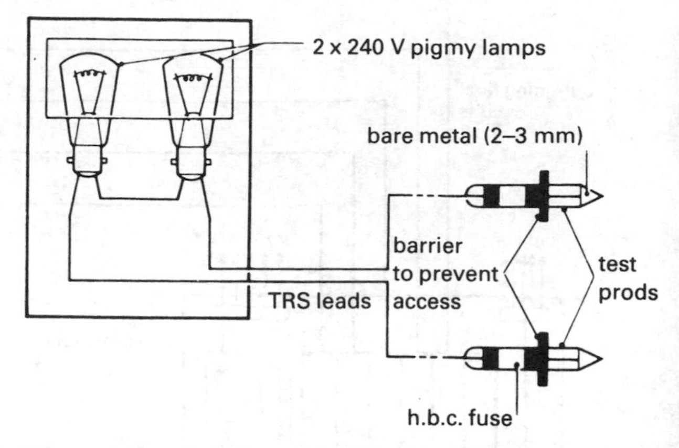

Figure 6.8 *Test lamp set*

Earth fault loop impedance tests are required under Reg. 713–10. The testers used for this are called 'phase-earth loop testers'. Analogue and digital testers are available with dual-scale readings of between 0–2Ω and 0–100Ω. They give a direct reading in ohms (remember impedance is measured in ohms). It should be noted that tables of the IEE Regulations should be used for comparing the results of circuits tested. The instruments switch a low value of resistance into the circuit being tested (usually 10Ω) and simulate a fault by injecting a test current of up to 25 A for a short duration of about 30–50 ms. Digital models are mostly used today, with their easier and more accurate readings (the 'Megger' LT5 is typical). Once values of the earth loop are obtained they can easily be converted into measurement values of a possible prospective short circuit. On this point, there are instruments available that are used for measuring short-circuit currents and fault levels up to 20 kA.

Figure 6.11 shows how an earth fault loop impedance test is made on a socket outlet. A supply is required for this and care must be taken to ensure that no ill-effects arise due to a defective earthing circuit.

A test is also required on the operating performance of installed *residual current devices* (Reg. 713–12); a supply is again required. Testers for this have to simulate a fault on the wiring installation in order to establish that the rated trip current of the circuit breaker operates within 0.2 s (or a delay time specified by the manufacturer of the tester). Such instruments are now made with a digital scale and have to record a tripping time of 40 ms when they are used for the purpose of reducing the risk of direct contact. Figure 6.12 shows a typical test circuit. In practice, several tests can be applied. If the device is of 30 mA sensitivity and used for shock risk protection then it should be:

(a) tried at half its tripping current (15 mA) and should not operate,
(b) tried at full tripping current (30 mA) and should operate within 0.2 s,
(c) tried at 150 mA and should operate within 40 ms.

It should also be tried quarterly by the consumer, once it has been installed.

Completion and inspection certificates

When an electrical installation has been finally inspected and tested in accordance with the IEE Regulations, completion and inspection certificates are required to be given to the customer (or client who had the work done) by the electrical contractor (or person carrying out the work); see Section 741. Any defects or omissions revealed will have to be made good before the completion certificate is issued. Appendix 6 of the IEE Regulations provides

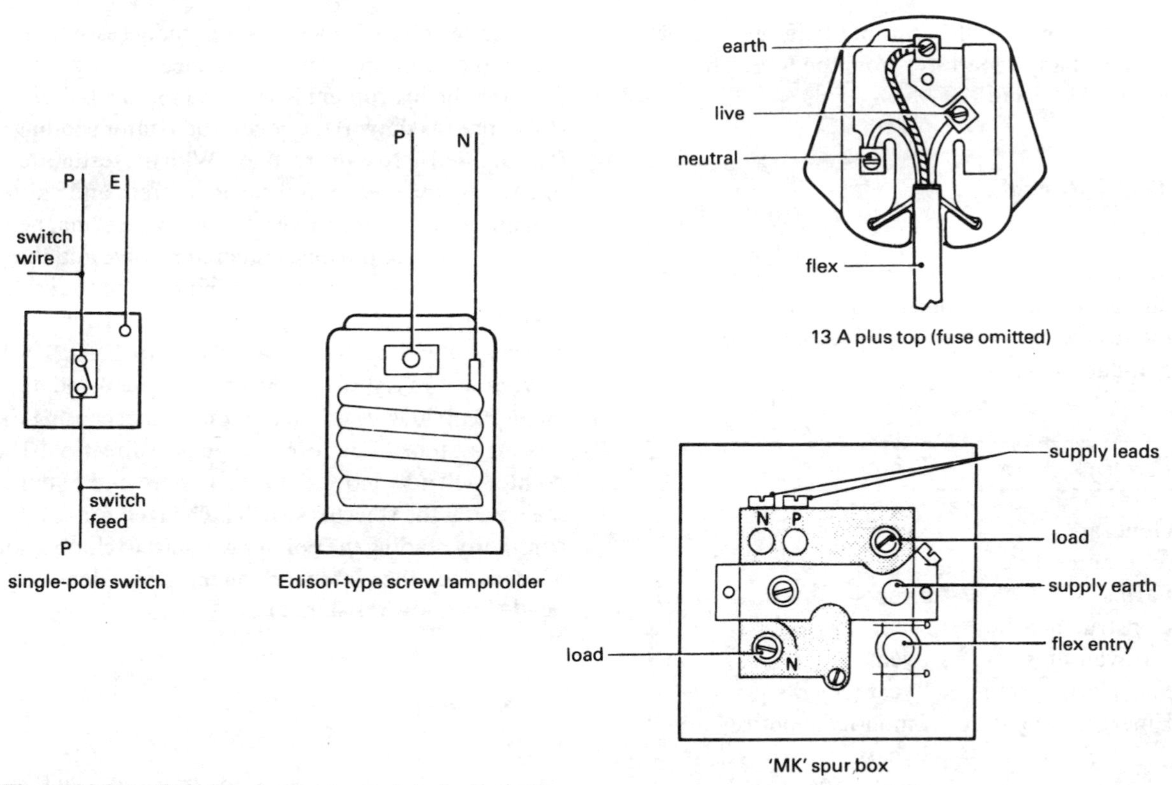

Figure 6.9 *Polarity connections: (a) single-pole switch; (b) Edison-type screw lampholder; (c) 13 A plug top (fuse omitted); (d) 'MK' spur box*

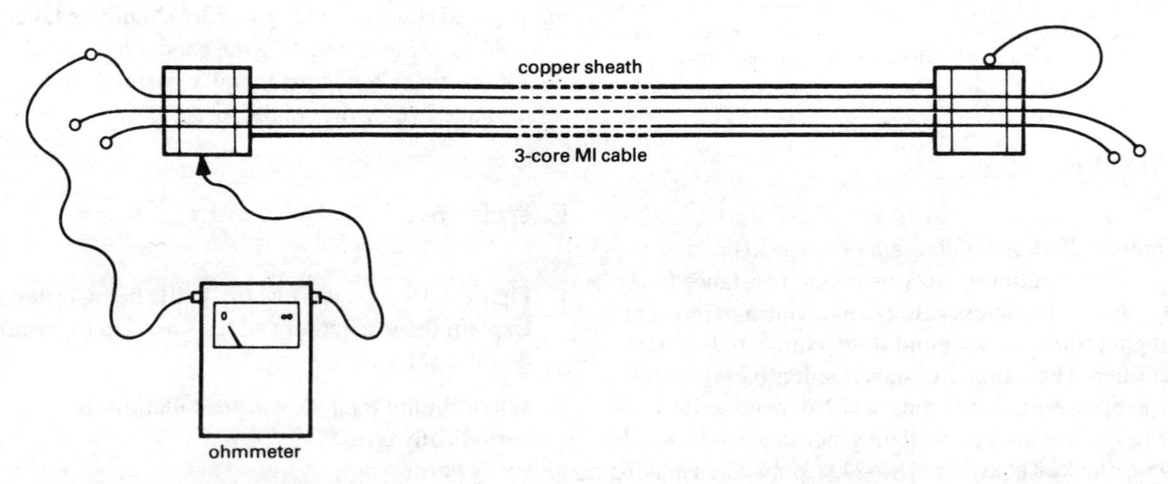

Figure 6.10 *Testing MI cable cores*

information on both forms of certificate and includes details of any departures from the Regulations (see Regs 120–04 and 120–05).

Test intervals

The intervals between inspection and testing of different types of premises is given in the table below; the test periods will either be recommended (R) or mandatory (M).

Premises	*Period*	*Test*
General (e.g. domestic)	5 years	R
Farms	3 years	R
Construction sites	3 months	R
Caravan sites	1 year	R
Fire-alarm systems	1 year (see BS 5839)	R
Emergency lighting	1 month/6 months/ 3 years (see BS 5266 Part 1)	R
Cinemas	1 year	M
Petrol-filling stations	1 year	M
Laundrettes	1 year	M
Churches	2 years (new) 1 year (old)	R
Factories	Regular intervals	M

Note: Some of the recommended tests may become mandatory if they are incorporated in local authority conditions of license.

The ohmmeter

Figure 6.13 shows a diagram of a typical hand-operated, continuity and insulation resistance tester. It is an a.c. brushless generator having a permanent magnet rotor and a wound stator connected to a static rectifier. The ohmmeter movement is based on the ratio of currents in two coils, which depend on the value of resistance under test. The generator handle needs to be cranked at around 184–240 r.p.m. when making tests. In the circuit diagram of Figure 6.14 there is a selector switch for either low-continuity resistance readings or high-insulation resistance readings.

When the instrument is switched for insulation resistance (as shown) the generator's stator winding is fully applied across the rectifier. With the tester's lead open, only the control coil carries current and its magnetic field interacts with the permanent magnetic field to cause the moving system to revolve and the pointer to move towards the infinity end of the scale (denoting open circuit, high resistance). When, however, the tester's leads are short-circuited, current will pass through the deflecting coil and its magnetic field will swamp the control coil causing the moving system to rotate in the opposite direction. The pointer will now move towards the zero end of this scale and if the selector switch is changed for continuity reading the pointer will start to climb again towards the other zero mark on this scale (denoting a dead-short, low resistance).

Tong tester

Figure 6.15 shows a diagram of a Crompton clip-on ammeter known as a tong tester. It is for use on a.c. or d.c. supplies and allows measurement of current without breaking the circuit. It is a light, robust and trouble-free instrument and it can be left in circuit without overheating. The jaws of the instrument are metal while the rest of its body is made from an insulating material. When using the instrument on uninsulated (bare) conductors, care should be taken to avoid earthing or shorting the conductor; it is also important not to touch the metallic parts of the tong tester made live by the conductor.

Exercise 6

1 Figure 6.16 shows an electric kettle being tested. Explain the inspection and tests needed to ensure that it is safe.

2 When should the following installations be periodically tested?
   (i) Farm
   (ii) Petrol-filling station

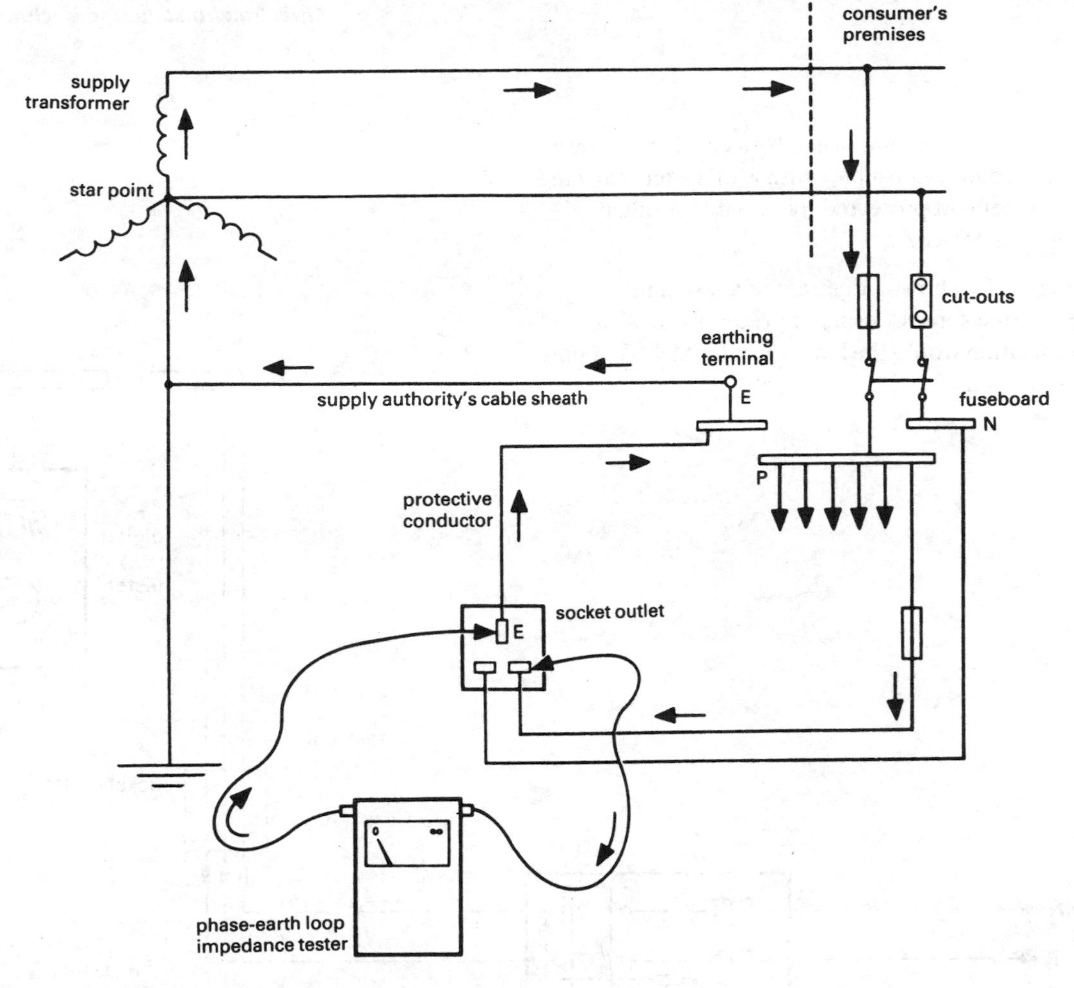

Figure 6.11 *Testing earth loop impedance in a TN–S system*

(iii) Construction site
(iv) Domestic dwelling
(v) Caravan site

3 Show, with the aid of a diagram, how a test is made for polarity on an Edison-type lampholder.

4 With regard to Figure 6.10 explain how you would test the cores of the MI cable for polarity, continuity and insulation resistance.

5 Explain how you would set about testing the stator windings of a three-phase induction *motor* for continuity and insulation resistance.

6 Determine the *maximum* earth fault loop impedance values for the following final circuits and, in each case, state the required current to cause disconnection:

(a) A 240 V immersion heater protected by a 15 A BS 3036 fuse,
(b) A 240 V motor protected by a 32 A h.b.c. BS 88 Part 2 fuse,
(c) A 240 V ring final circuit protected by a 30 A BS 3871 m.c.b.

7 A factory's electrical wiring was subdivided for insulation resistance tests. The various sections produced readings of: 40 MΩ, 20 MΩ, 1 MΩ, 5 MΩ and 100 MΩ. If the test had been done once, at the main intake point, what equivalent resistance value would this produce?

8 What are the IEE Wiring Regulations concerning completion certificates and inspection certificates?

9 What is the maximum earth electrode resistance allowed for a consumer with a TT system earthing arrangement protected by a 30 mA residual current device?

10 Figure 6.8 shows a diagram of a test lamp set. Describe some of its main precautions. What indication would the lamps give on a 240 V supply compared with a 415 V supply?

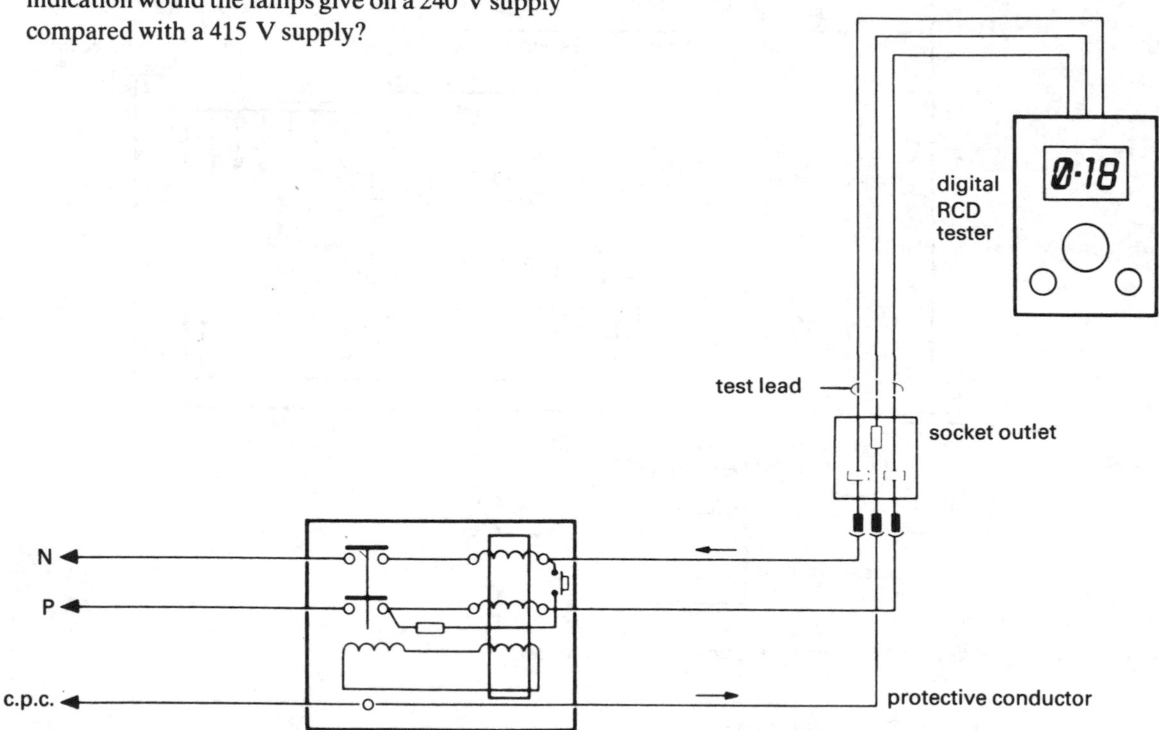

Figure 6.12 *Test on a residual current device*

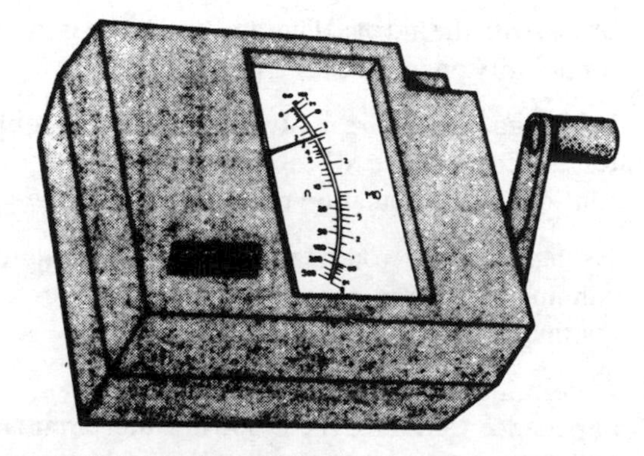

Figure 6.13 *'Wee Megger' continuity and insulation resistance tester*

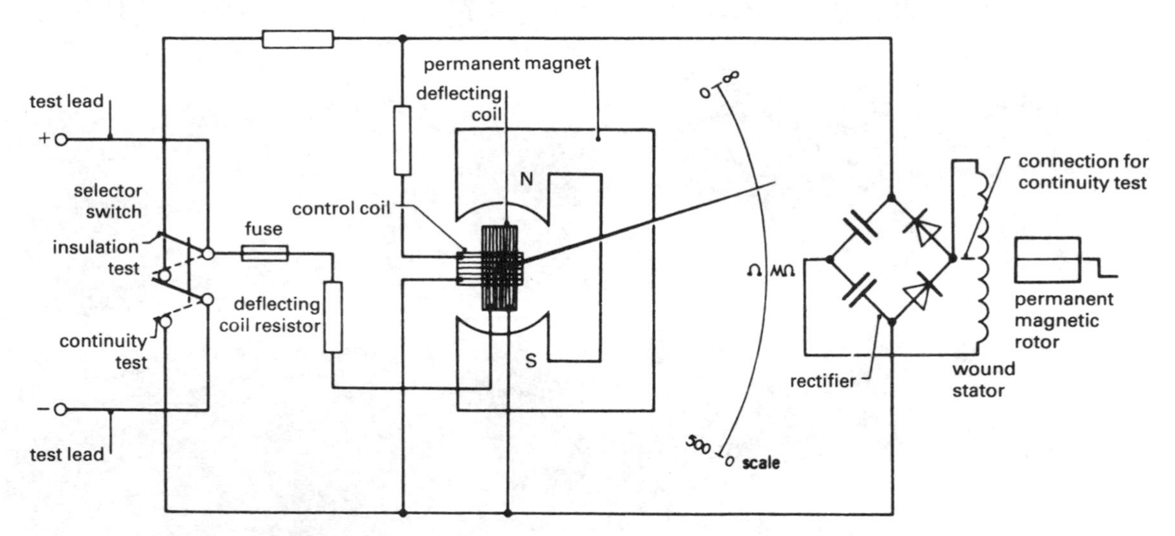

Figure 6.14 *Circuit diagram of ohmmeter tester*

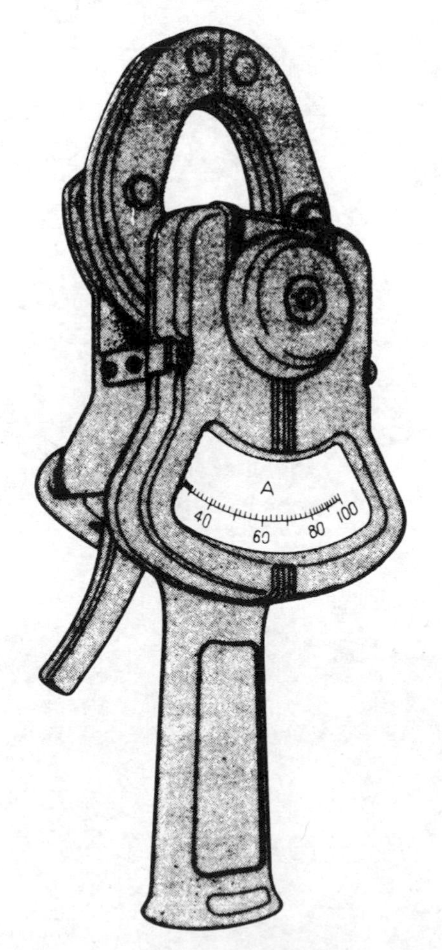

Figure 6.15 *a.c./d.c. tong test clip-on ammeter*

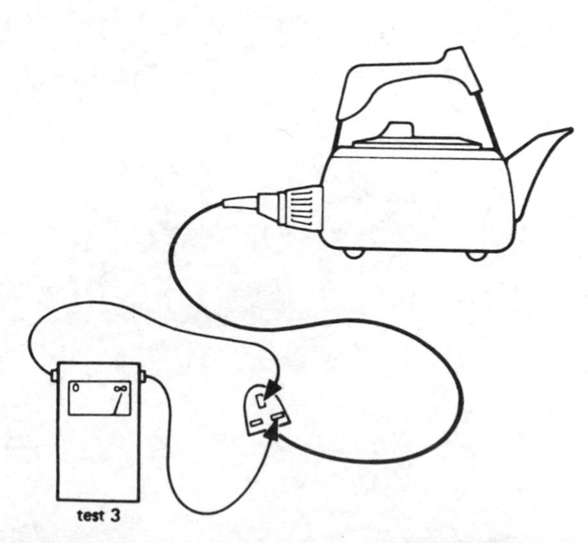

Figure 6.16 *Testing insulation resistance of a kettle*

chapter seven

Questions and answers

Multiple-choice question papers

Part I Certificate

1 The line diagram shown in Figure 7.1 represents the control circuit for
(a) three-gang, two-way switching
(b) two-way, four-way switching
(c) two-way intermediate switching
(d) double-pole, two-way switching

2 With reference to Figure 7.2 which hands-touching situation is likely to prove the most serious?
(a) between an unbroken neutral
(b) between yellow and blue phases
(c) between red phase and neutral
(d) between an unbroken red

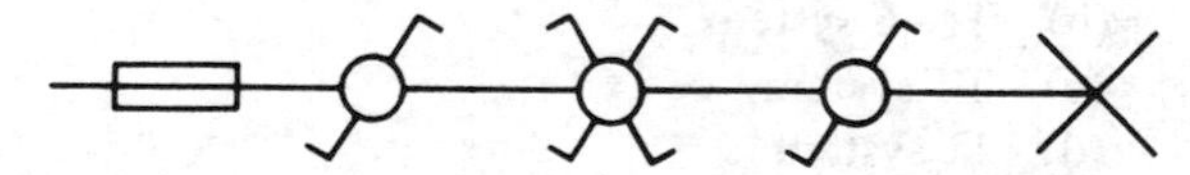

Figure 7.1

3 Which method of earthing is no longer recognised today?
(a) mains water pipe
(b) supply cable sheath/armouring
(c) copper rod electrode
(d) metallic reinforcement in concrete

4 In a fluorescent lamp, the purpose of the *choke* is to
(a) improve power factor
(b) absorb ultraviolet radiation
(c) improve tube life
(d) limit supply current

5 An emergency lighting circuit is classified as
(a) Cateogry 0
(b) Category 1
(c) Category 2
(d) Category 3

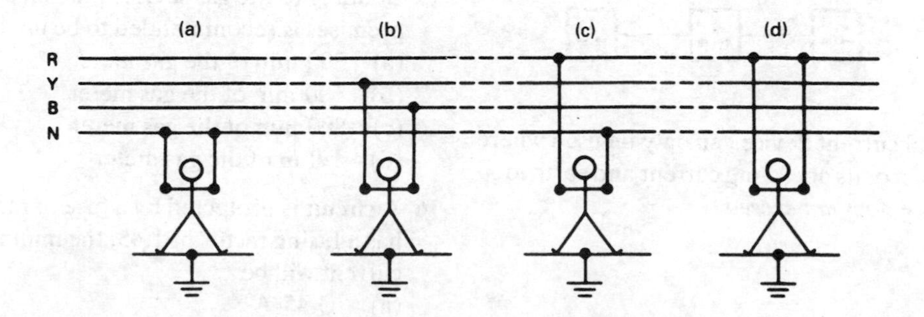

Figure 7.2

6 All phase colours of a three-core flexible cable should be identified as
(a) red
(b) brown
(c) blue
(d) black

7 Which of the following circuit protective devices is known as a semi-enclosed rewireable fuse?
(a) BS 88
(b) BS 1361
(c) BS 3036
(d) BS 3871

8 A 12 kW/240 V cooking appliance has a control point incorporating a 13 A socket outlet. The assumed current demand (see IEE Regulations) using a 54% diversity allowance is
(a) 21 A
(b) 27 A
(c) 30 A
(d) 45 A

9 Figure 7.3 shows a 30 A ring final circuit in PVC conduit. How many single-core PVC cables will be run in section XY?
(a) three
(b) four
(c) six
(d) nine

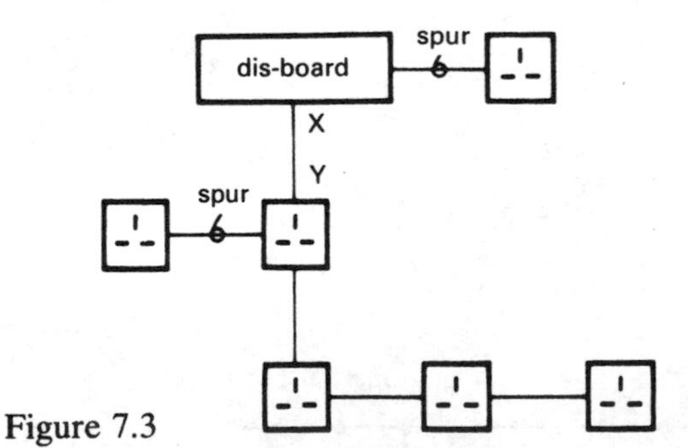

Figure 7.3

10 A residual current device can only be used where the product of its operating current and earth loop impedance *does not exceed*
(a) 25
(b) 30
(c) 45
(d) 50

11 Unless otherwise stated by a British Standard, the insulation resistance between live conductors and exposed conductive framework of an electrical appliance should not be less than
(a) 0.5Ω
(b) 1.0Ω
(c) 0.5 MΩ
(d) 1.0 MΩ

12 The name of the BS 3939 graphical location symbol shown in Figure 7.4 is
(a) intake point
(b) energy meter
(c) distribution board
(d) contactor

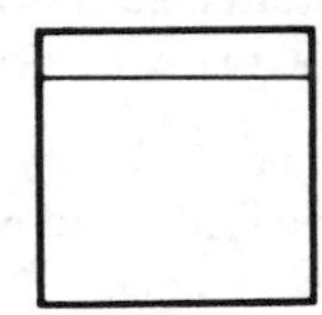
Figure 7.4

13 Which earthing system uses a PEN conductor?
(a) TN–C system
(b) TN–S system
(c) TT system
(d) IT system

14 All phase and neutral conductors of a.c. circuits shall be contained within the same ferrous enclosures to avoid
(a) wiring difficulties
(b) eddy currents
(c) excessive voltage drop
(d) circuit overloading

15 Bonding to the gas service pipe on a consumer's premises is recommended to be made within
(a) 500 mm of the gas meter
(b) 600 mm of the gas meter
(c) 800 mm of the gas meter
(d) 1.0 m of the gas meter

16 A circuit is protected by a fuse of rating 5 A. If it has a fusing factor of 1.45, the minimum fusing current will be
(a) 3.45 A
(b) 6.25 A

(c) 7.25 A
(d) 10.00 A

17 The term '*spur*' used in electrical definitions means
(a) branch cable
(b) outlet-box
(c) ring main box
(d) circuit flex

18 The correction factor associated with a BS 3036 fuse is
(a) 0.500
(b) 0.725
(c) 1.500
(d) 2.000

19 Under fault conditions, protective devices controlling fixed equipment circuits have to be disconnected within a time of
(a) 0.4 s
(b) 0.5 s
(c) 4.0 s
(d) 5.0 s

20 For outdoor use, it is recommended to use a residual current device that trips out at
(a) 30 mA
(b) 100 mA
(c) 300 mA
(d) 500 mA

21 Materials delivered to a site should be checked against the delivery note and also the
(a) job sheet
(b) time sheet
(c) original order
(d) variation order

22 The term '*short-circuit current*' is used to explain an overcurrent condition that flows in a circuit and
(a) leaves it electrically sound
(b) causes a residual current device to operate
(c) results in a fault of negligible impedance
(d) only allows h.b.c. fuses to operate

23 The permissible voltage drop allowed in a consumer's installations must not exceed
(a) 2.5%
(b) 4.0%
(c) 6.0%
(d) 7.5%

24 In Figure 7.5, the recommended distance allowed between the control point and cooking appliance is
(a) 1 m
(b) 2 m
(c) 3 m
(d) 4 m

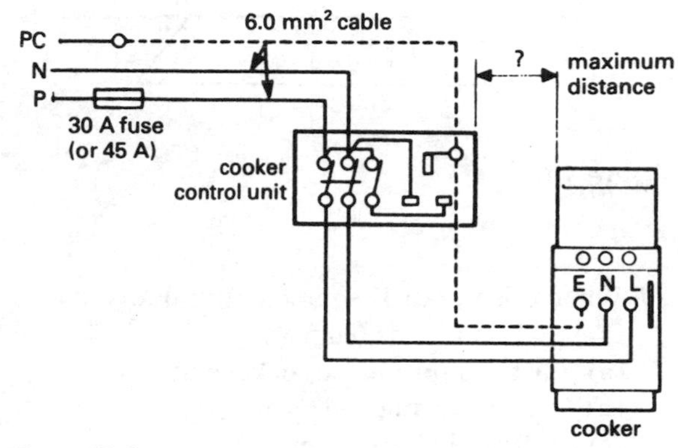

Figure 7.5

25 The IEE Wiring Regulations require residual current devices to be tested
(a) weekly
(b) monthly
(c) quarterly
(d) yearly

26 The minimum depth for drilling holes in a wooden joist in order to run cables is
(a) 15 mm
(b) 25 mm
(c) 35 mm
(d) 50 mm

27 The minimum size of supplementary bonding permitted in a bathroom where no mechanical protection is provided is
(a) 1.0 mm^2
(b) 1.5 mm^2
(c) 2.5 mm^2
(d) 4.0 mm^2

28 In Figure 7.6, the minimum height for the overhead cable is
(a) 3.5 m
(b) 4.2 m

(c) 4.8 m
(d) 5.8 m

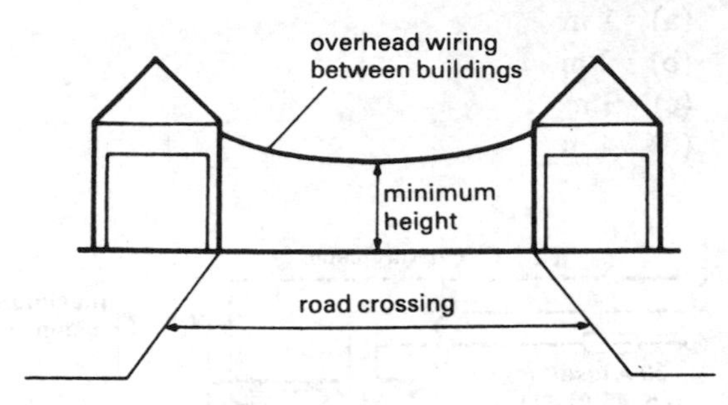

Figure 7.6

29 When using Class II equipment outdoors, it is essential for it to be fed from
(a) an r.c.d.-protected socket outlet
(b) a circuit protected by an m.c.b.
(c) a TT earthing system
(d) an SPN switch

30 The method of bonding items of metalwork together is to
(a) avoid the flow of neutral currents
(b) ensure a common potential exists
(c) safeguard against corrosion
(d) safeguard against short-circuit faults

31 When using an Edison screw-type lampholder, the centre contact should be connected to the
(a) c.p.c. conductor
(b) neutral conductor
(c) switch conductor
(d) supply conductor

32 The compression ring used in a brass MI cable gland is to
(a) provide mechanical protection
(b) prevent the ingress of moisture
(c) provide earth continuity
(d) prevent misalignment

33 The recommended interval between inspection tests for a farm installation is
(a) 2 years
(b) 3 years
(c) 4 years
(d) 5 years

34 When a fault occurs at Point X in Figure 7.7, discrimination of operation between fuse A and fuse B is achieved only if
(a) fuse A and fuse B rupture together
(b) fuse A ruptures leaving fuse B intact
(c) fuse B ruptures leaving fuse A intact
(d) fuse B ruptures with 5 s

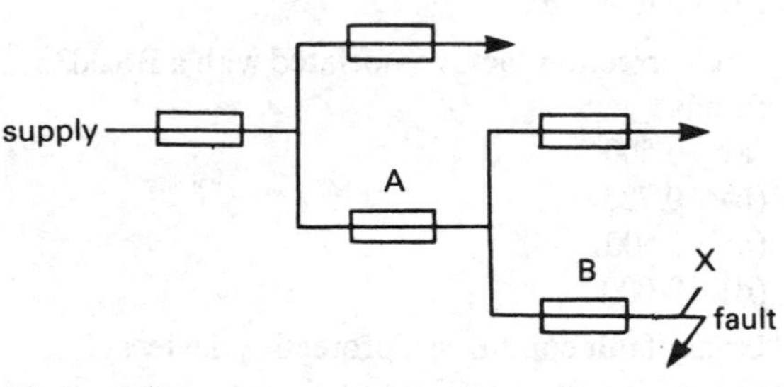

Figure 7.7

35 Where a building is being constructed, equipment which is easily damaged should be installed in position.
(a) as soon as the circuit wiring is complete
(b) as soon as it is delivered on site
(c) with a caution notice attached
(d) when other work operations are complete

36 M.i.m.s. cable is sometimes provided with an overall covering of PVC to
(a) lower its ambient temperature
(b) prevent corrosion of its sheath
(c) protect it from sunlight
(d) increase its insulating property

37 Which diagram in Figure 7.8 shows the correct connection of an Edison screw lampholder?

38 Magnesium oxide is a material that acts as
(a) an insulator
(b) an electrolyte
(c) a semiconductor
(d) a conductor

39 Figure 7.9 shows a diagram of a tungsten filament lamp. The code used for this lamp is
(a) SOX
(b) MCF
(c) SON
(d) GLS

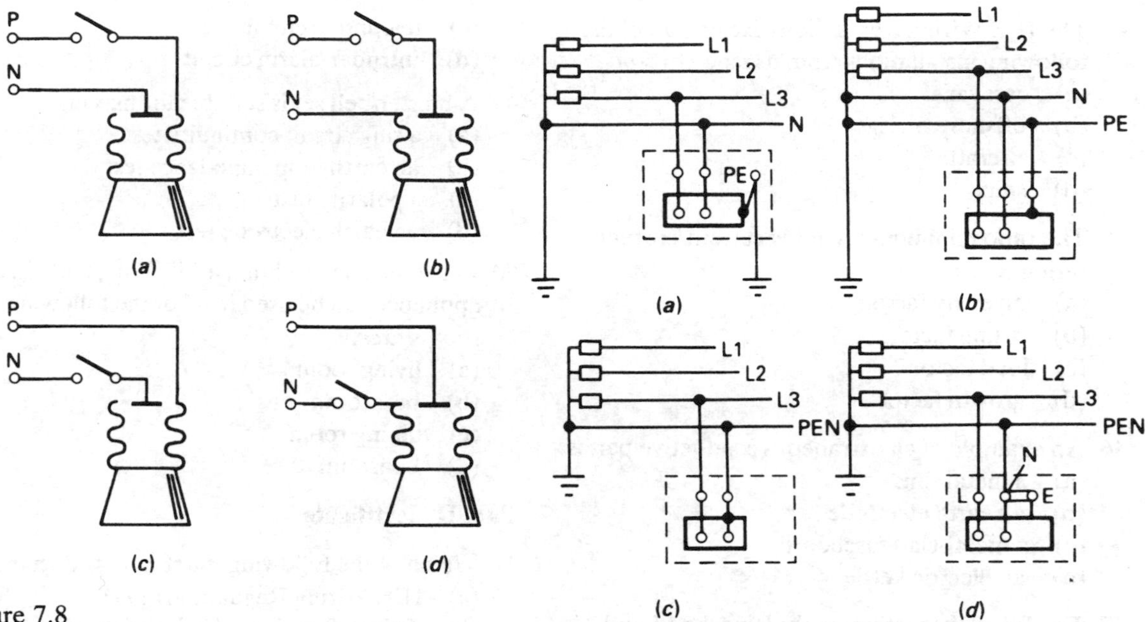

Figure 7.8

Figure 7.10

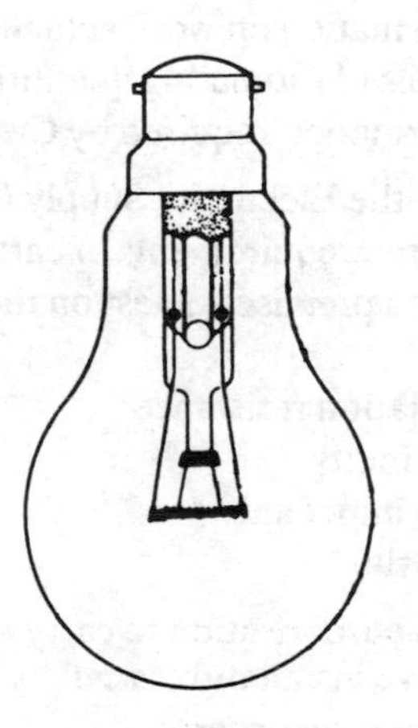

Figure 7.9

40 For household premises, the floor area served by 13 A BS 1363 socket outlets on a ring final circuit *should not exceed*
(a) 20 m^2
(b) 50 m^2
(c) 100 m^2
(d) 200 m^2

41 Which of the distribution earthing systems in Figure 7.10 is called a TT system?

42 Which of the following protective devices has a thermal overload and magnetic overload characteristic?
(a) rewireable fuse to BS 3036
(b) cartridge fuse to BS 1361
(c) cartridge fuse to BS 1362
(d) miniature circuit breaker to BS 3871

43 The symbol shown in Figure 7.11 is marked on tested equipment to denote use in
(a) explosive areas
(b) exhibition halls
(c) external conditions
(d) extra circuits

Figure 7.11

44 The IEE Wiring Regulations exclude all of the following installations from its scope *except*
(a) caravans
(b) offshore oil rigs
(c) aircraft
(d) ships

45 The ratio minimum breaking current/current rating is called
(a) diversity factor
(b) fusing factor
(c) load factor
(d) growth factor

46 An example of an extraneous conductive part is
(a) a metal sink
(b) an earth electrode
(c) a metal-clad fuseboard
(d) an electric kettle

47 The BS 3939 location symbols marked 1 and 2 in Figure 7.12 are the responsibility of the
(a) Department of Environment
(b) generating company
(c) Regional Electricity Company
(d) electrical consumer

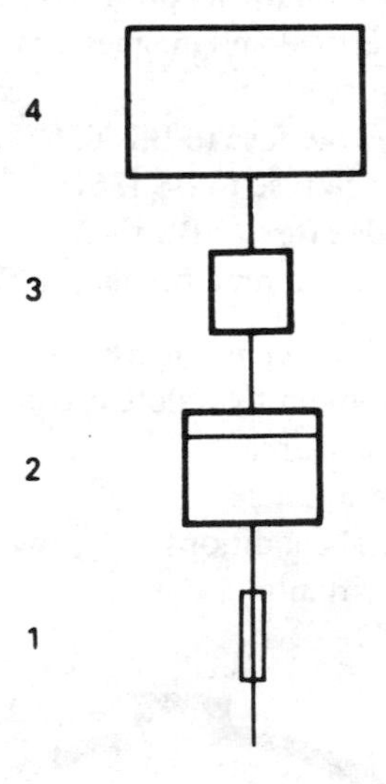

Figure 7.12

48 Which of the following is associated with a Category 3 circuit?
(a) low-voltage lighting circuit
(b) fire-alarm circuit
(c) telephone circuit
(d) intruder-alarm circuit

49 A buzzer/bell set is suitable for making
(a) a ring circuit continuity test
(b) an earth loop impedance test
(c) a polarity test
(d) an earth electrode test

50 In a domestic dwelling, a 240 V portable heating appliance can be used in all of the following rooms *except*
(a) living room
(b) bathroom
(c) dining room
(d) bedroom

Part II Certificate

1 Which of the following is a statutory document?
(a) IEE Wiring Regulations 1991
(b) British Standards Codes of Practice
(c) Health and Safety at Work Act 1974
(d) BS 5266 Emergency Lighting of Premises

2 The object of an 'as-fitted' drawing is to provide
(a) work programmes for site operatives
(b) information on work actually done on site
(c) details of estimating quantities
(d) information required by Clerks of Works

3 To satisfy the Electricity Supply Regulations of 1988 and the requirements of earth leakage on a consumer's premises, a test on the wiring is made for
(a) insulation resistance
(b) continuity
(c) loop impedance
(d) polarity

4 A written authorisation to carry out work on or about a live installation is called
(a) a variation order
(b) a completion certificate
(c) a permit to work
(d) an inspection certificate

5 The BS 3939 graphical symbol shown in Figure 7.13 is an
(a) illuminated sign
(b) emergency lighting point
(c) indicator panel
(d) automatic fire detector

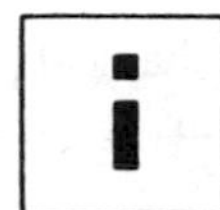

Figure 7.13

6 A sodium lamp designated the code SOX has an arc tube that operates at
(a) high pressure
(b) medium pressure
(c) low pressure
(d) extra-low pressure

7 The purpose of a flameproof piece of equipment is to
(a) allow it to be used at high temperatures
(b) contain within it any internal explosion
(c) prevent an internal explosion
(d) function normally in the event of a fire

8 When PVC cables are used for both Category 1 and Category 3 circuits, and installed in a common trunking, they must
(a) be run next to each other
(b) be segregated by fireproof barriers
(c) operate at the same voltage
(d) come from the same distribution board

9 When a dangerous situation is discovered on site, the first action to be taken by an electrical trainee is to
(a) write a note to the building-site foreman
(b) place a danger notice nearby
(c) report the matter to his electrician
(d) leave it to be discovered by a site safety officer

10 Which of the following installations is excluded from the IEE Wiring Regulations?
(a) caravan sites
(b) construction sites
(c) farm installations
(d) offshore installations

11 What is the minimum current demand for the installation in Figure 7.14, if consideration is given to the diversity allowances (in brackets) for each circuit item?
(a) 66 A
(b) 77 A
(c) 85 A
(d) 175 A

12 The estimated maximum external impedance (Z_E) for a TN–C–S supply earthing system is
(a) 0.35Ω
(b) 0.80Ω
(c) 1.45Ω
(d) 2.00Ω

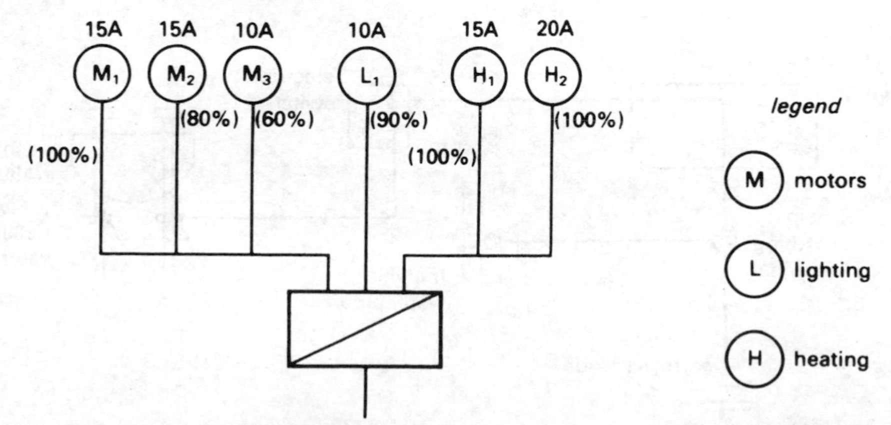

Figure 7.14

13 What is the current demand for six twin 80 W, 240 V fluorescent luminaires applying the 1.8 factor?
(a) 2.0 A
(b) 3.6 A
(c) 4.0 A
(d) 7.2 A

14 An installation is earthed in order to
(a) protect it against short-circuit faults
(b) enable polarity to be defined
(c) prevent exposed conductive parts becoming live
(d) prevent voltages appearing on the neutral

15 Which of the following is not suitable for isolation purposes?
(a) circuit breaker
(b) thyristor
(c) plug and socket
(d) fuse link

16 In Figure 7.15, which of the following is a circuit protective conductor?
(a) 1
(b) 2
(c) 3
(d) 4

17 In Figure 7.15 which of the following is a supplementary bonding conductor?
(a) 1
(b) 2
(c) 3
(d) 4

18 Supplementary bonding conductors that are provided with mechanical protection need not be of a size larger than
(a) 1.5 mm^2
(b) 2.5 mm^2
(c) 4.0 mm^2
(d) 6.0 mm^2

19 The circuit protective device supplying a radiant heater in a bathroom should, if a fault occurred, disconnect the circuit in
(a) 300 ms
(b) 400 ms
(c) 500 ms
(d) 600 ms

20 The stationary part of an induction motor is called the
(a) cage
(b) stator
(c) armature
(d) commutator

21 Which of the following motors *does not* have a commutator?
(a) universal motor
(b) shunt motor
(c) compound motor
(d) cage induction motor

22 The letters to denote a cage induction motor's main stator windings are:
(a) U1 and U2
(b) F1 and F2
(c) W1 and W2
(d) Z1 and Z2

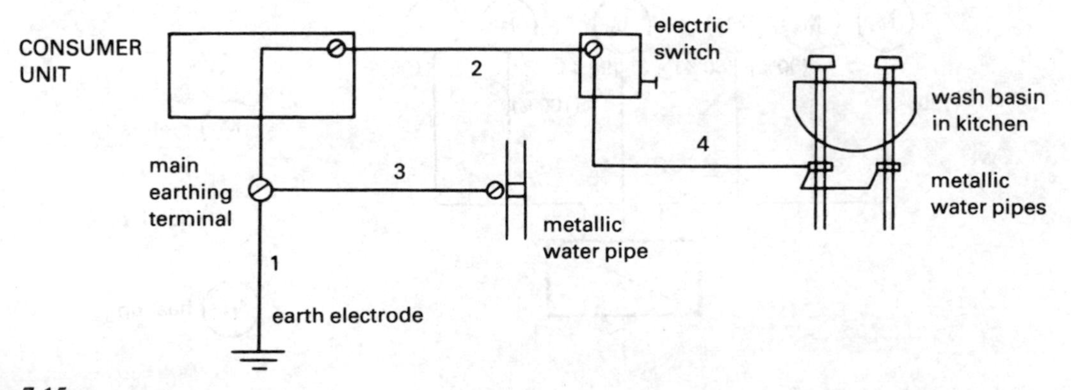

Figure 7.15

23 Which of the following is not a standard a.c. supply voltage?
(a) 240 V
(b) 415 V
(c) 11 kV
(d) 32 kV

24 Which of the following types of a.c. starter reduces the supply voltage to approximately 58% at starting?
(a) direct on-line
(b) autotransformer
(c) star-delta
(d) rotor-resistance

25 For remote control of a direct on-line a.c. contactor starter, start buttons are wired in
(a) series and stop buttons are wired in parallel
(b) parallel and stop buttons are wired in series
(c) series and stop buttons are wired in series
(d) parallel and stop buttons are wired in parallel

26 Which of the following lamps has the lowest efficacy?
(a) tungsten filament lamp
(b) fluorescent tube
(c) SON discharge lamp
(d) MBF/U mercury lamp

27 In a connected fluorescent lamp switch-start circuit, a short occurs in the glow-type starter causing the tube to
(a) glow brightly at both ends
(b) flicker on and off
(c) fail to light
(d) flicker intermittently

28 From the time-current characteristics shown in Figure 7.16, the curves above the 7 s line denote the circuit breaker's ability to operate on
(a) normal overload
(b) short circuit
(c) open circuit
(d) mechanical shock

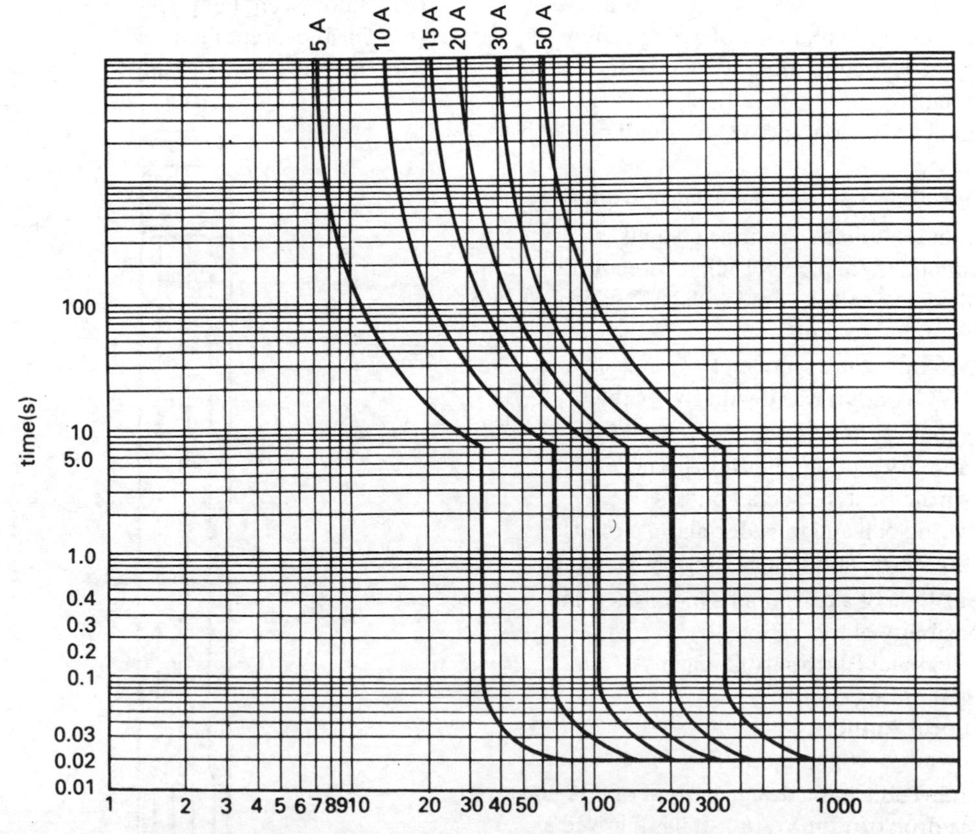

Figure 7.16 prospective current (A) Type 2 m.c.b. time-current characteristics

29 For an overcurrent of 30 A in Figure 7.16, which device will trip first?
  (a) 5 A
  (b) 10 A
  (c) 15 A
  (d) 20 A

30 From Figure 7.16 which m.c.b. will operate at 800 A?
  (a) 5 A device only
  (b) 5 A and 10 A device only
  (c) 5 A, 10 A and 15 A devices only
  (d) any of the devices

31 The effects of an a.c. motor's stator winding on an open circuit fault is called:
  (a) armature reaction
  (b) single-phasing
  (c) split-phasing
  (d) hunting

32 The purpose of a circuit chart found inside a distribution board is to provide information about the
  (a) circuit route and type of wiring system
  (b) type of circuit protective device and earthing facility
  (c) name of circuit and size of circuit protective device
  (d) electrical contractor's name and address

33 Which of the following wiring systems is recommended for a petrol-filling forecourt?
  (a) flexible conduit wired with PVC cables
  (b) trunking system
  (c) M.i.m.s. cables having PVC oversheaths
  (d) PVC conduit wired with PVC cables

34 A ring main system is one that is used for
  (a) primary a.c. supplies into premises
  (b) wiring BS 1363 socket outlets
  (c) wiring bell and intruder-alarm circuits
  (d) fire-alarm and other Category 3 circuits

35 The earthing of a consumer's premises is the responsibility of the
  (a) Regional Electricity Company
  (b) generating company
  (c) Local Authority
  (d) owner/occupier

36 A certain cable has a design current of 8 A and a voltage drop of 15 mV/A/m. If it is allowed a maximum 4% declared voltage drop, the longest length that can be used is
  (a) 25 m
  (b) 33 m
  (c) 75 m
  (d) 100 m

37 Which of the following is *not* recognised as a suitable earth electrode?
  (a) non-corrosive metallic covering of cables
  (b) metallic reinforcement in concrete
  (c) electrode embedded in a foundation
  (d) metal gas and water pipes

38 For a completed electrical installation, the minimum insulation resistance test-reading allowed is
  (a) 0.5 MΩ
  (b) 1.0 MΩ
  (c) 100 MΩ
  (d) infinity

39 The type of lamp shown in Figure 7.17 is called a
  (a) sodium lamp
  (b) fluorescent lamp
  (c) incandescent lamp
  (d) mercury-vapour lamp

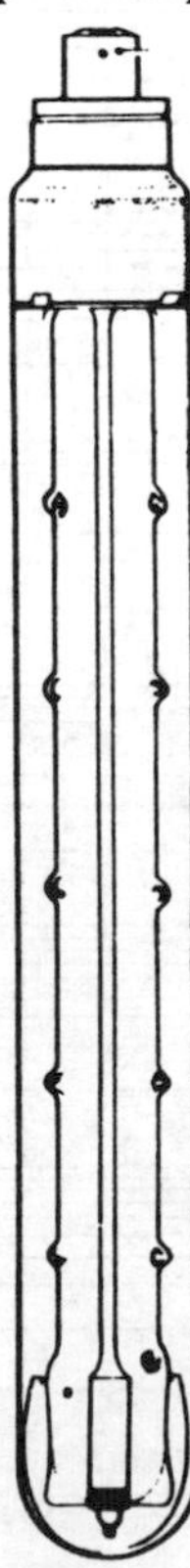

Figure 7.17

40 In a flameproof installation, a conduit stopper box should be fitted
(a) only at termination points in danger areas
(b) every 3 m length of run through the system
(c) only on the side of the wiring remote from danger
(d) only on the side classified as a danger zone

41 Figure 7.18 shows a radial distributor and each section AB, BC and CD has a resistance of 0.01Ω. What is the voltage at point C?
(a) 225 V
(b) 230 V
(c) 235 V
(d) 240 V

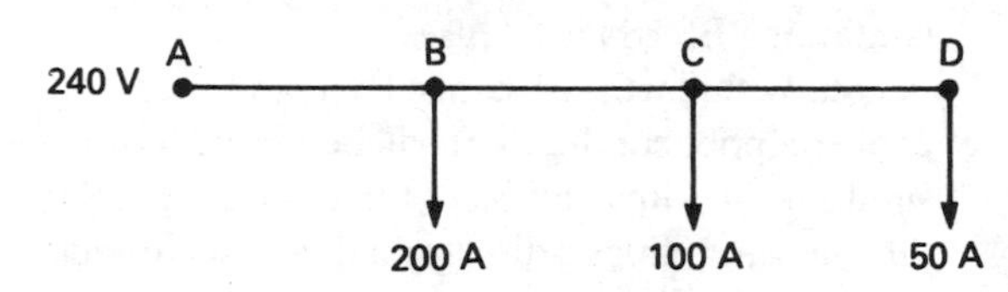

Figure 7.18

42 The maximum earth loop impedance for a 240 V a.c. fixed equipment final circuit protected by a 10 A Type 2 m.c.b (see Figure 7.16) is
(a) 2.4Ω
(b) 3.4Ω
(c) 5.0Ω
(d) 6.0Ω

43 A tong tester is an instrument used to measure
(a) frequency
(b) vibration
(c) power
(d) current

44 Which set of the circuit connections in Figure 7.19 will reverse the motor's direction?
(a) P and N
(b) S and T
(c) C1 and C2
(d) U1 and U2

45 One of the main reasons for using a ring main to feed a large distribution system is
(a) supply reliability
(b) spare standby switchgear
(c) consumer growth factor
(d) less maintenance

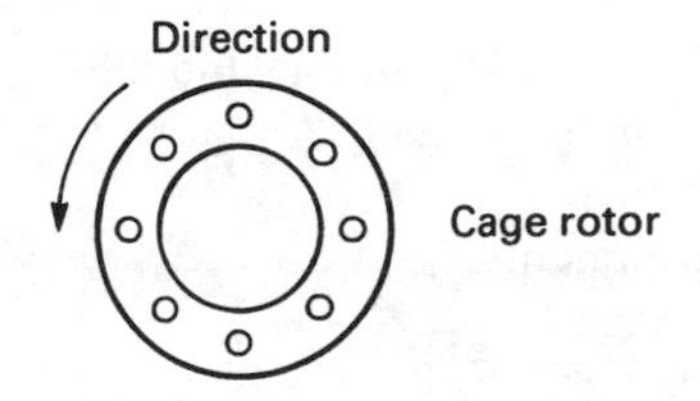

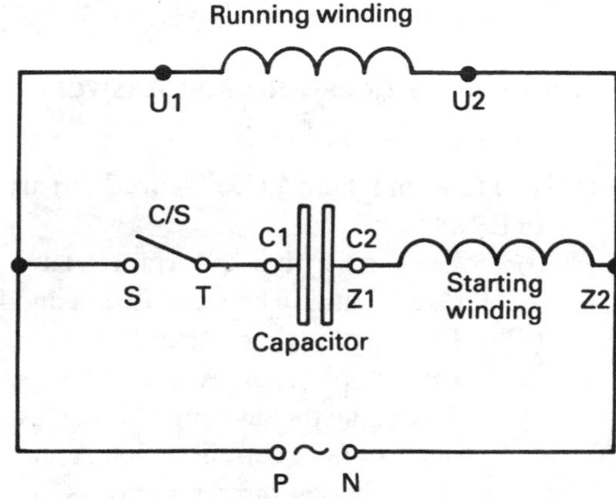

Figure 7.19

46 Which of the following instruments measures energy?
(a) integrating meter
(b) wattmeter
(c) power factor meter
(d) frequency meter

47 Which of the following switches can be used for emergency switching?
(a) micrograp switch
(b) semiconductor switch
(c) fireman's switch
(d) interlocking switch

48 To reduce stroboscopic effects, discharge lamps may be connected
(a) on one phase of a three-phase supply
(b) on all three phases of a three-phase supply
(c) without capacitor power factor improvement
(d) with special filters

49 If the phase conductor of a supply system is 50 mm², the minimum size of circuit protective conductor is according to Table 54G IEE Regs.
(a) 10 mm²
(b) 16 mm²
(c) 25 mm²
(d) 35 mm²

50 The formula $S = \frac{\sqrt{I^2 t}}{k}$ is used for finding the safe size of a protective conductor against
(a) thermal constraints
(b) corrosive environments
(c) mechanical damage
(d) circuit grouping

City and Guilds questions and answers

1 (a) The fuses in Figure 1 (see Figure 7.20) are to BS 88 Part 2.
(i) Explain why it is important to have discrimination between fuse A and B.
(ii) By using the time-current characteristics of fuses A and B, determine the disconnection times due to a fault current of 300 A and state whether discrimination is achieved.
(b) State why certain types of miniature circuit breaker have both thermal and magnetic tripping devices.

CGLI Part II June 1987

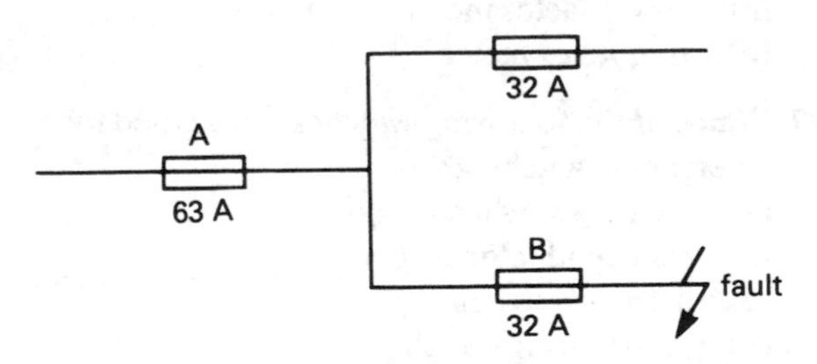

Figure 7.20

Solution

1 (a) (i) Discrimination may be defined as the ability of a circuit protective device to interrupt the supply to a faulty circuit without interfering with the source of supply to remaining healthy circuits in the system.
(ii) From Figure 8, IEE Regulations, the 32 A fuse disconnects in 0.12 s. From Figure 9, IEE Regulations, the 63 A fuse disconnects in 4.0 s. It will be seen that fuse discrimination is achieved with the 32 A fuse operating first.
(b) The m.c.b. is designed with the ability to protect circuits against two overcurrent conditions, namely *overload* and *short circuit*. Overloads are detected by a thermal bimetal element, whereas short circuits are detected by a solenoid operating instantaneously.

2 An electric heater rated at 3 kW/240 V is to be installed as a fixed appliance. The feed will be from an existing spareway in a fuseboard containing BS 88 Part 2 fuses.
A twin-with-earth, PVC-insulated and sheathed cable (copper conductors) will be run with three similar cables in an ambient temperature of 25°C, bunched and clipped direct and totally surrounded by thermal insulation for most of its 12 m run. The grouping factor shall be selected from Table 4B1 and the voltage drop of this circuit must not exceed 6 V.
(a) (i) Determine the design current (I_b) and the nominal rating of the fuse (I_n).
(ii) Calculate the minimum current rating of the required cable.
(iii) State the table and columns used in the IEE Regulations to determine the initial size of the cable.
(iv) Select the minimum size of cable to comply with the requirements for current-carrying capacity and voltage drop.
(v) Calculate the actual voltage drop.
(b) State *two* further requirements that have to be met before a final cable size can be determined.

CGLI Part II June 1987

Solution

2 (a) (i) The design current
$I_b = P/V = 3000/240 = 12.5$ A.
From Table 41A2 (a) of the IEE Regulations, $I_n = 16$ A

(ii) The minimum current rating of the required cable will have to consider three correction factors, one for grouping C_g, one for thermal insulation C_i and one for ambient temperature C_a.
$C_g = 0.65$ from Table 4B1 and
$C_a = 0.5$
from Note of Reg. 523–04
$C_a = 1.03$ from Table 4C1.
Thus: $I_z = I_n$/correction factors
$= 16/0.65 \times 0.5 \times 1.03$
$= 47.79$ A.

(iii) From the IEE Regulations, Table 4D2A and columns 1 and 6 (Method 1) are used.

(iv) A 10 mm^2 will be selected.

(v) The cable selected has an I_r value of 63 A and a voltage drop/A/m of 4.4. Since $V = L \times I_b \times$ mV/A/m, the cable voltage drop is:
$V = 12 \times 12.5 \times 0.0044 = 0.66$ V.

(b) One is *disconnection time* and the other is *thermal constraint.*

3 Consumers' switchgear should conform to the IEE Regulations requirements for isolation, switching (functional, emergency and mechanical maintenance), control and protection against overload and short circuit.

(a) Figure 1 (Figure 7.21) shows the equipment for operating a group of motors. Using the numbers indicated as identification, state the function, or functions, performed by *each* item of equipment with reference to isolation, switching, control and protection.

(b) State the requirements of the Regulations regarding the use of semiconductor devices for isolation, switching and control.

CGLI Part II June 1986

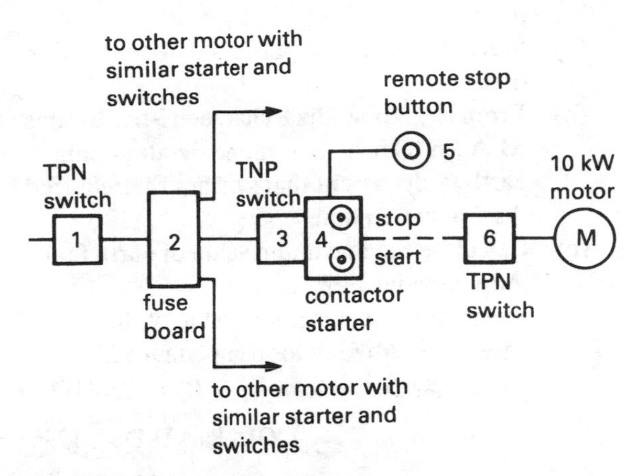

Figure 7.21

Solution

3 (a) 1 The TPN switch is the first most important means of control for the motor circuit, its purpose being to provide *isolation* at the origin of the circuit. It is an 'off-load' device.

2 The fuseboard will provide the means of *excess current protection* to all the circuits indicated. This protection will include h.b.c. fuse back-up protection for the local TPN switches and starters in (3), as well as overcurrent protection of the circuit cables.

3 This will provide the motor with its own isolation and may also serve the purpose of *switching off for mechanical maintenance*. It has to be manually operated and have reliable indication of its 'on' and 'off' functions, as an 'on-load' device should have locking-off facilities.

4 The starter will provide *functional switching* as well as no-volt and overcurrent protection of the motor and the cables between the starter and motor.

5 The remote stop buttons will provide *emergency switching*; these have to be of the 'stay closed' type or 'latched' type, and be coloured red.

6 This switch provides local *isolation* for maintenance, such as the removal of the motor, and it should be provided with locking-off facilities.

(b) See IEE Regulations, Reg. 537–02–04.

4 A 10 mm^2, two-core PVCSWA cable, supplying 240 V equipment outside the main equipotential zone, is protected by a 63 A BS 88 fuse that also gives protection against indirect contact.

(a) State the maximum permitted disconnection time.
(b) From the Appendix 8 characteristics for this 63 A BS 88 fuse, determine the minimum earth fault current that satisfies the required disconnection time.
(c) Calculate the maximum value of earth fault loop impedance Z_s.
(d) Determine the maximum value of the external earth fault loop impedance Z_E, given that the value of $R_1 + R_2 = 0.144\Omega$

CGLI Part II Dec. 1986

Solution

4 (a) The answer to this is found in Reg. 471–08–03 (ii), i.e. 0.4 s (Table 41A)
(b) From the time–current characteristics, the fuse will require 500 A
(c) From Appendix 8 and transposing the formula for I_F, then

$$Z_s = \frac{U_o}{I_F}$$
$$= 240/500 = 0.48\Omega$$

(d) Since $Z_s = Z_E + R_1 + R_2$
then $Z_E = Z_s - (R_1 + R_2)$
$= 0.48 - 0.114$
$= 0.366\Omega$

5 (a) With the aid of a circuit diagram, explain the method of starting a three-phase, wound rotor (slip ring) induction motor.
(b) Figure 2 (Figure 7.22) shows a fixed-length conductor with a magnet moving past it in the direction shown.
(i) In which direction would the induced e.m.f. in the conductor act?
(ii) State *two* factors upon which the magnitude of the e.m.f. depends.

CGLI Part II June 1987

Solution

5 (a) A diagram of this method of starting is shown in Figure 7.23. The majority of medium size slip-ring motors are started by connecting

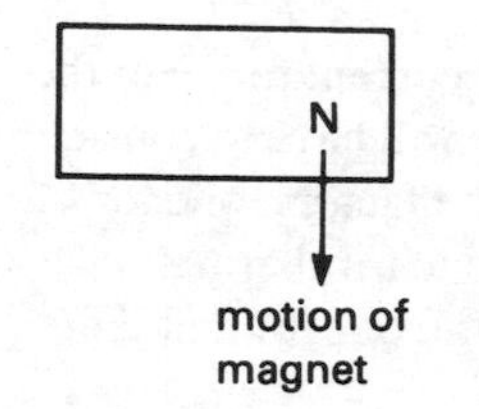

Figure 7.22

their stator windings directly to the three-phase a.c. supply via a push button, direct on-line contactor starter. The starting current and the starting torque are controlled by an external resistance connected to the rotor circuit across slip rings. This resistance is progressively cut out as the motor speeds up, until finally the rotor winding is short-circuited. In practice, the starter is fitted with interlocks to ensure that the resistance is all-in when it is first started. Large motors are often started with a liquid resistance, which provides closer control over the starting current and a much smoother start.

(b) (i) This answer is found by applying Fleming's right-hand rule to Figure 7.22. The direction of current in the conductor will be represented as a dot (current coming out).
(ii) Two factors are: *strength of magnetic flux* and *velocity of the magnetic field cutting the conductor at right angles.*

N.B. Students should know the expression: $e = Blv$ volts.

6 During the course of the work on a construction site, an electrician falls from a ladder and appears to be unconscious.
Detail the immediate and subsequent actions to be taken by:
(a) the person finding him
(b) the person in charge of the site.

CGLI Part II June 1984

Solution

6 (a) The question is very open and it implies that the person finding the electrician assumes, firstly, that he has fallen from the ladder.

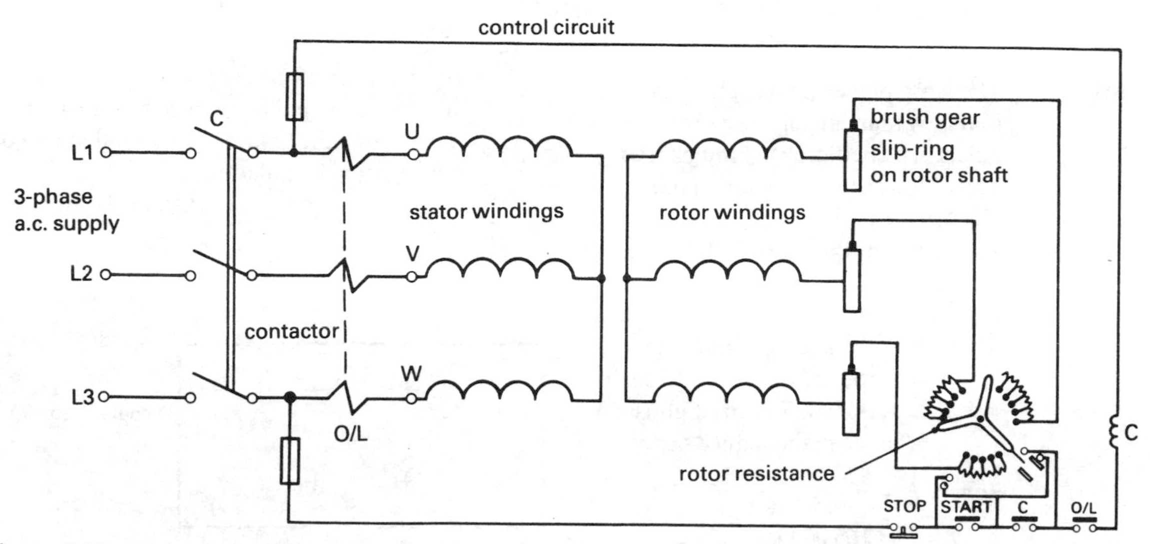

Figure 7.23 *Rotor resistance starting for a slip-ring induction motor*

This assumption may of course be wrong, but the actions to be taken are as follows.

(i) *Check for signs of breathing* by placing your ear over the electrician's mouth and listen as well as look along the chest for signs of movement.

(ii) If the electrician is breathing, check to see if there are any *signs of bleeding*. If not, keep him warm with additional clothing—this might be your own coat or jacket.

(iii) *Do not attempt to move him* in case he has other injuries.

(iv) *Shout for assistance.*

(v) If the electrician does not seem to be breathing, get someone to call medical assistance while you begin *mouth-to-mouth ventilation.*

Note: Instructions on this method are given in Figure 2.13.

(vi) It is important that someone stays with the injured person until medical help arrives.

(b) The person in charge of the site will be informed of the accident and he may himself have initiated the call for an ambulance or doctor. He may also have a person on site trained in first-aid treatment who will be able to give him details additional to those obtained from the person discovering the accident. Once the injured person has been treated or taken to hospital and the next of kin informed, a report of the accident has to be written for the employer, as well as an accident report on the lines described in Figure 2.15. The accident may need investigating by a safety officer so the area should be left as it is and items involved should not be removed.

7 (a) The premises of a shop measuring 15 m × 8 m are to be illuminated by 240 V tungsten filament lamps having an efficacy of 16.5 lumen per watt. The shop is to have 20 luminaires and an average illuminance of 550 lux. Assuming a utilisation factor of 0.5 and a maintenance factor of 0.8, calculate

(i) the power in watts per luminaire

(ii) the total power

(iii) the total current.

(b) (i) State the precautions that should be taken at the luminaires in (a) to comply with IEE Regulations 523–1 and 523–3.

(ii) Assuming that the luminaire lampholders in (a) are of type B22, state how many circuits should be used for the lighting.

CGLI Part II Dec. 1985

Solution

7 (a) (i) The lamp power for each luminaire is found by rearranging the expression of efficacy, i.e. efficacy = lumens/watt. Thus: Power (watt) = lumen output/efficacy
The lumen output (F) of all the lamps is found by the expression:
$F = E \times A/C \times M$
where E is the average illuminance
A is the area required to be lit
C is the coefficient of utilisation
M is the maintenance factor

Hence: $F = \dfrac{550 \times 15 \times 8}{0.5 \times 0.8}$
$= 165\,000$ lm

Since there are twenty luminaires, each one will need to produce 165 000/20 = 8250 1m. Since the efficacy is given, the power in watts per luminaire is:
$P = 8250/16.5$
$= 500$ W

(ii) The total power will be 500 × 20 = 10 000 W

(iii) The total current is P/V = 10 000/240
= 41.67 A

(b) (i) These Regulations refer to the high temperatures likely to be encountered inside luminaires and call for the use of appropriate heat-resistant insulation or sleeving—see Table 10B, IEE Regulations (15th Edition).

(ii) Reference should be made to Reg. 553–14 and Table 55B. For a B22 lampholder, the maximum rating of the protective device is 16 A. As each luminaire takes 2.08 A, this means only 7.69 (actually 7) can be used for each circuit.
Hence, *three circuits* will be needed.

8 Figure 3 (Figure 7.24) shows the lighting plan of a room drawn to a scale of 1:100. Each luminaire is of the discharge type and has a rating of 40 W and an efficacy of 100 lm/W. The utilisation and maintenance factors are 0.5 and 0.8, respectively.

(a) Calculate:
(i) the area of the room
(ii) the average illuminance in the room.
(b) Determine the assumed current demand of this lighting installation.

CGLI Part II June 1987

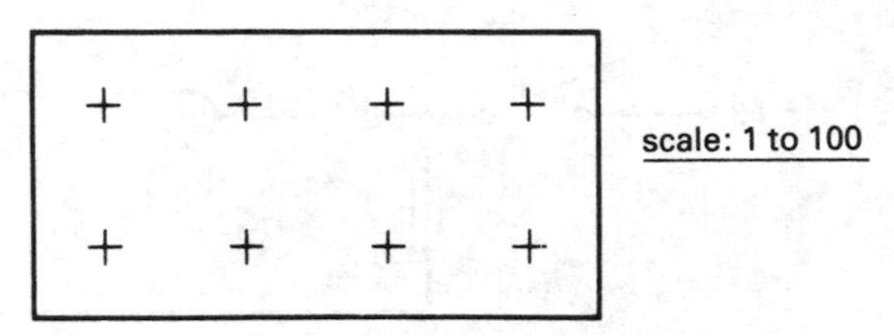

Figure 7.24 *Lighting plan*

Solution

8 (a) (i) A = 32 m^2
(ii) $E = F \times C \times M/A$ (see Question 7 above)
$= (4000 \times 8) \times 0.5 \times 0.8/32$
$= 400$ lux

(b) $I = P \times 1.8/V$
$= 8 \times 40 \times 1.8/240$
$= 2.4$ A

9 (a) Explain why it is usual to:
(i) use high voltage for transmission and distribution of electrical energy
(ii) balance loads on a three-phase system.
(b) Draw a simple circuit diagram showing how an 11 kV, three-phase, three-wire supply is transformed down to provide a 415 V, three-phase, four-wire supply. Indicate the line and phase voltages.
(c) Draw a simple circuit diagram to show how the following supplies can be obtained from the same three-phase, four-wire a.c. system
(i) 415 V, three-phase, three-wire
(ii) 415 V, single-phase
(iii) 240 V, single-phase.

CGLI Part II Dec. 1986

Solution

9 (a) (i) Transformers provide a very convenient method of changing voltage from one pressure to another. Raising the voltage automatically reduces current and it is this advantage that allows great amounts of power to be transmitted over large distances

(ii) The object of balancing loads over three-phases of the supply is to create an even current demand on each phase. By doing this, circuit cables and switchgear can be properly designed and selected.

(b) See Figure 7.25

(c) See Figure 7.25

10 Briefly explain the application of the following types of emergency lighting:

(a) escape lighting
(b) safety lighting
(c) standby lighting.

Solution

10 (a) Escape lighting is where the level of illuminance needs to be maintained at all times to ensure adequate illuminance throughout all escape routes—the minimum is 0.2 lux.

(b) Safety lighting is where the level of illuminance needs to be maintained at such a level that there is no risk to persons involved in potentially hazardous processes, should the normal lighting fail.

(c) Standby lighting is lighting that must be of sufficient illuminance to carry on normal activities on the failure of public supply mains—the level of lighting depends upon the activity being performed.

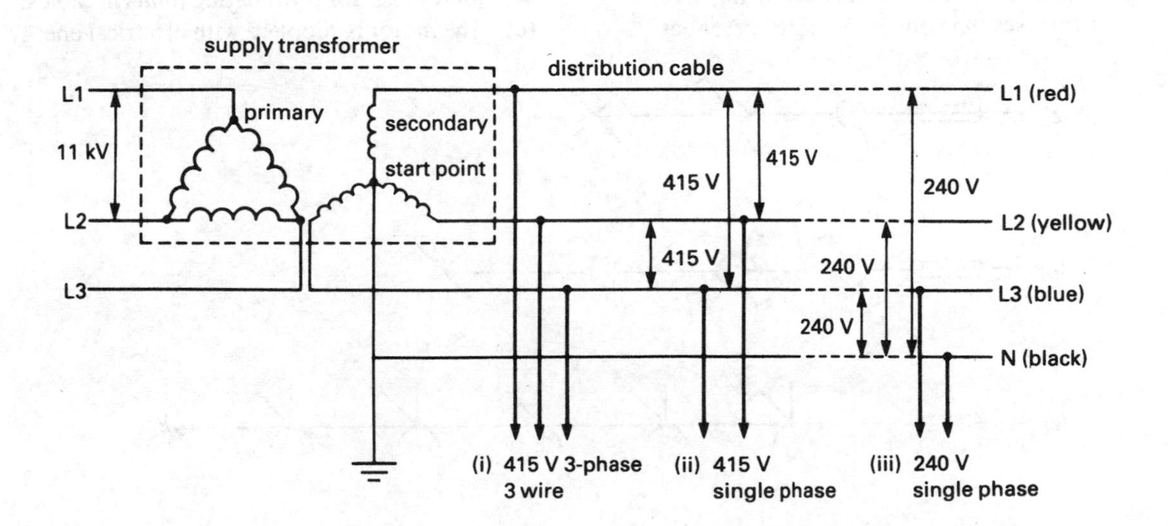

Figure 7.25 *Standard supply voltages*

Answers to exercises

Exercise 1

1 See useful definitions (pages 1–4).
2 See Figure 7.26
3 See useful definitions. For example
  (a) *open circuit* This generally refers to an electrical circuit that has become broken or discontinuous so current cannot flow.
  (b) *closed circuit* This generally refers to an electrical circuit that is continuous so that current will flow.
  (c) *short circuit* This generally refers to a fault condition whereby live conductors are shorted out.
  (d) *polarity* In electrical terms this is an indication of conductor polarity, whether it is a 'phase', 'neutral', or 'earth' conductor or whether it is of 'positive' or 'negative' polarity.
  (e) *conductor resistance* is the resistance of the material that allows the current to flow and has a low ohmic value.
  (f) *insulation resistance* The insulation medium surrounding and supporting a live or potentially live conductor. Its ohmic value should be very high.
  (g) *fusing factor* This is a factor which expresses the minimum fusing current at which a fuse element will melt divided by the current rating.
  (h) *space factor* This is a ratio of the sum of the overall cross-sectional areas of cables (including sheath) to the internal cross-sectional area of their enclosure such as conduit and trunking.
  (i) *continuity* In electrical terms, this implies a cable's ability to conduct a continuous flow of current through circuit conductors. Earth continuity implies an uninterrupted path, preferably of low resistance.
  (j) *emergency switching* is the rapid cutting off of electrical energy to remove a hazard.

4 See pages 55–56 and Figure 4.9.
5 See Figure 7.27.
6 (a) U_o (b) I_p (c) I_z (d) I_2 (e) C_g (f) Z_s (g) R_1 (h) S (i) I_n (j) R_2.
7 The nominal setting of the protective device can be equal to or greater than the design current and equal to but not greater than the lowest current-carrying capacity of the circuit conductor.
8 (a) A switchfuse has stationary fuses, whereas a fused switch carries the fuses on its moving contacts.
  (b) A junction box is a conduit box, whereas a joint box is for terminating joints in cables.
  (c) The motor is supplied with electrical energy,

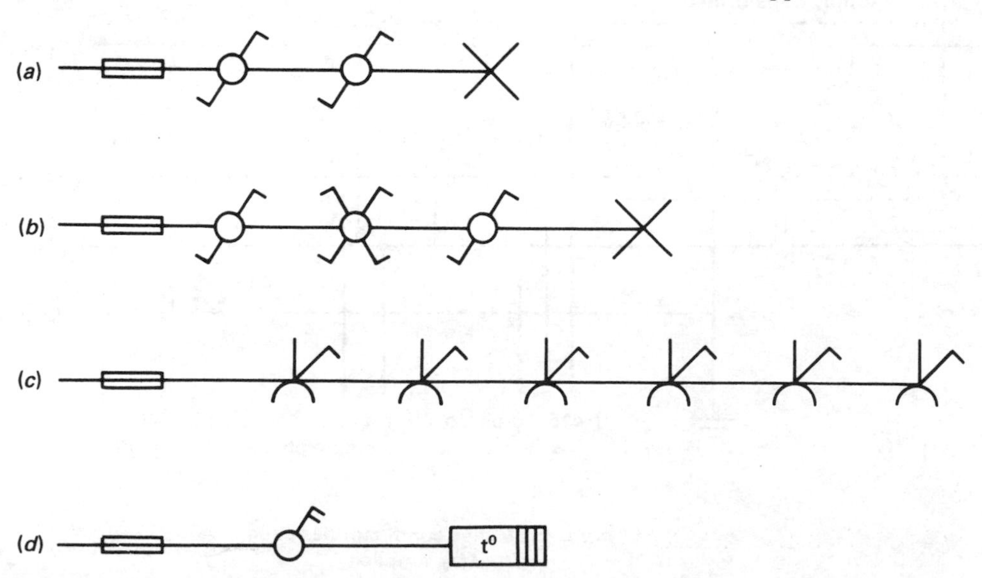

Figure 7.26 *Line diagrams using BS 3939 graphical location symbols*

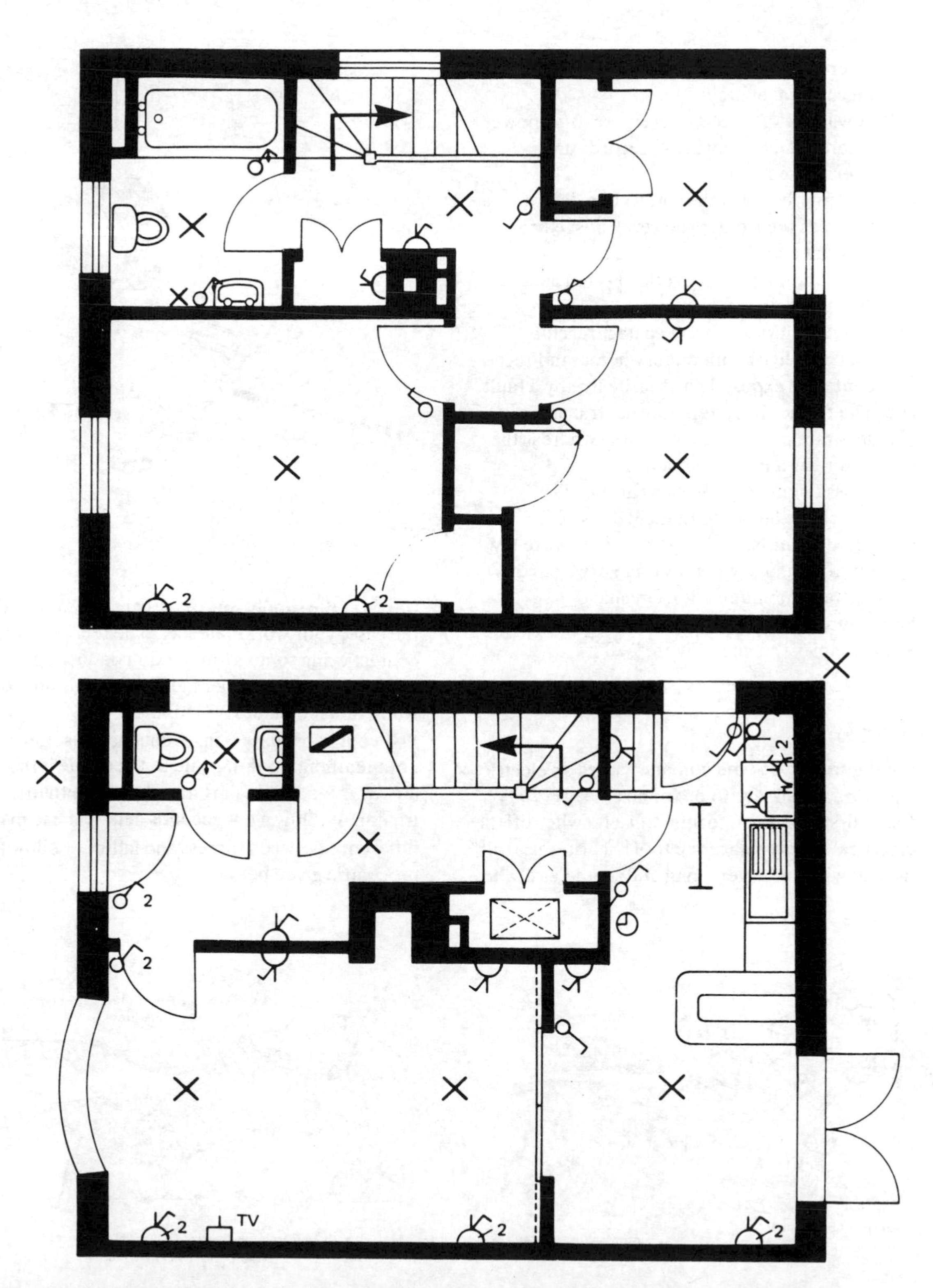

Figure 7.27 *Diagram showing use of BS 3939 graphical location symbols in a three-bedroomed house*

whereas the generator is supplied with mechanical energy.

(d) A wattmeter is an instrument reading power, whereas an energy meter records units of energy used.

(e) The m.c.b. is a circuit protective device, whereas the r.c.d. protects against earth leakage and electric shock risk.

(f) See definitions—an overload is not as severe as a short circuit.

(g) See definitions—direct contact is actually touching live conductors, whereas indirect contact is exposed parts made live by a fault.

(h) The former has a separate neutral and separate protective conductor, whereas the latter has a PEN conductor.

(i) Class I equipment has an earth facility, whereas Class II equipment does not.

(j) See definitions—Category 1 circuits are low-voltage mains, whereas Category 2 circuits are telecommunication circuits.

9 See Figure 7.12.

10 See examples shown in Figure 7.28.

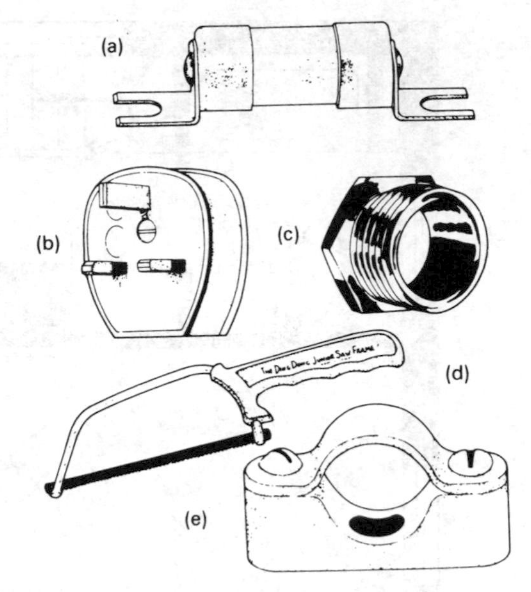
(a)
(b)
(c)
(d)
(e)

Figure 7.28

Exercise 2

1 See Figure 7.29. Some guidance notes on electric shock treatment are to be found on pages 26–27. One's first reaction is to immediately switch off the electrical supply. Since the tool is a portable drill its lead will not be very long and should easily be traced to the supply source. Where this may not be the case, your workmate has to be released from contact using some form of non-conducting material, such as rubber gloves, dry clothing, dry wood or a length of PVC tubing.

If your workmate is in an unconscious state and not breathing (looking pale or even blue if the airway is blocked), start mouth-to-mouth first aid treatment. Only a few seconds delay can mean the difference between success and failure. Follow the procedure given below.

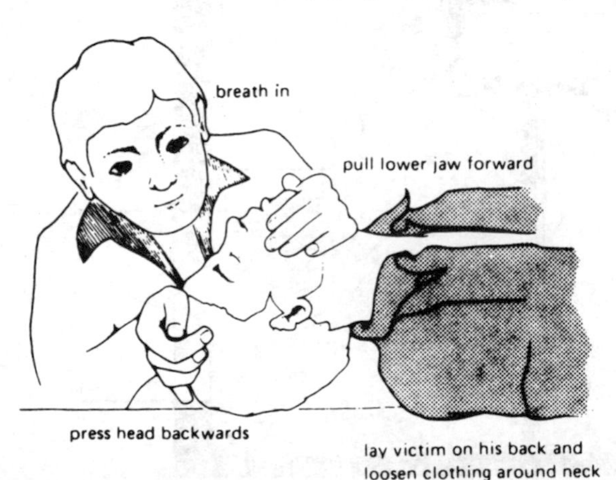
breath in
pull lower jaw forward
press head backwards
lay victim on his back and
loosen clothing around neck

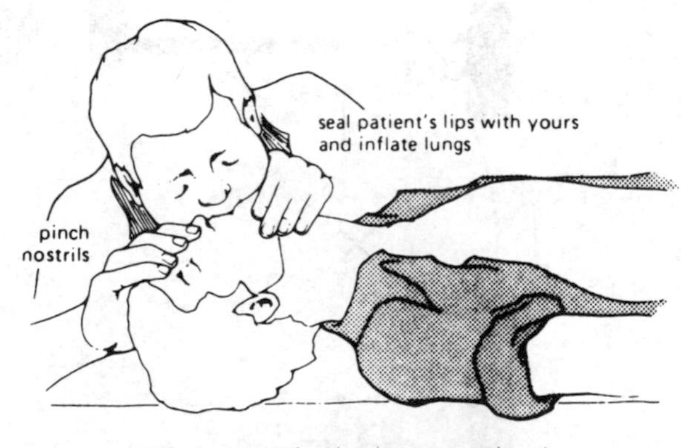
seal patient's lips with yours
and inflate lungs
pinch
nostrils
blow into lungs (twelve times every minute)
avoid patient's exhaled air

Figure 7.29 *Mouth-to-mouth ventilation*

NOTICE OF ACCIDENT OR DANGEROUS OCCURRENCE

1. OCCUPIER OF PREMISES
Name J.A.TOWNSHOTT
Address 5, APPLEBY WAY, ELY
Nature of Business BRICKWORKS

2. EMPLOYER OF INJURED PERSON (if different from above)
Name
Address

3. INJURED PERSON ~~Resident/Staff~~ ~~Widow/Widower~~ Married/~~Single~~
Surname SPARK
Christian Names IAN
Date of Birth 6.7.38 Occupation ELECTRICIAN
Address 12 OAKLANE WOOTEN
Name and address of parent or guardian
Mr. & Mrs. J.A. SPARK
108 ELM DRIVE, WOOTEN

4. PLACE WHERE INCIDENT OCCURRED
Address 5 APPLEBY WAY ELY
Exact Location (e.g. staircase to office, canteen, storeroom, classroom).
STOREROOM
Name of Person supervising J. FLASH

5. INJURIES AND DISABLEMENT
Fatal or non-fatal NON FATAL
Nature and extent of injury (e.g. fracture of leg, laceration of arm, scalded foot, scratch on hand followed by sepsis).
ELECTRIC SHOCK FROM FAULTY PORTABLE DRILL

6. ACCIDENT OR DANGEROUS OCCURRENCE
Date 13TH AUGUST 1988 Time 11-00am
Full details of how the incident occurred and what the injured person was doing.
If a fall of person or materials, plant, etc. state height of fall.
Mr. SPARK WAS DRILLING A METAL BOX AT THE TIME OF THE ACCIDENT
Name and address of any witness.
Mr. P. BROWN
22 WINDSOR ROAD, LILLY, KENT
If due to machinery, state name and type of machine.
What part of the machine caused the accident?
Was the machine in motion by mechanical power at the time?

7. ACTION FOLLOWING THE ACCIDENT
What happened? Mr. SPARK WAS TREATED FOR ELECTRIC SHOCK
When was the doctor informed? 11:00am
When did he attend? 11·15am
If taken to hospital, say when and where.
11-45 am UPTON HOSPITAL
Names and addresses of friends or relatives who have been notified of the accident:
PARENTS (SEE ABOVE)
When and how were they informed? 11-45am - TELEPHONE

Name of Doctor P.S. BONE
(address and telephone)
BROMHAM LANE SURGERY
BROM 61062

Signature of injured person or person completing this form:
J. Flash
Date: 13-8-83
If the form is completed by some person acting on the injured person's behalf, the address and occupation of such person should be entered.

Figure 7.30

(a) Kneel by your workmate's head and quickly inspect the mouth for any obstructions. Loosen clothing around the neck.
(b) Move the head fully back (as shown), breathe in deeply, seal your lips over his or hers, pinch the nostrils with one hand and breathe out into the lungs.
(c) Watch the chest rise and then turn your head away and breathe in once more.
Repeat the procedure about ten to twelve times every minute and continue until he or she is breathing satisfactorily or until you are told by a doctor to stop.
(d) If your workmate recovers before the doctor arrives keep him or her warm and place in a recovery position.

2 It refers to the voltage drop not causing the equipment to become damaged and if no BS requirement exists it must not exceed 4% of the nominal voltage from supply intake point to end of final circuits.
3 Follow the guidelines. A type of accident form is given in Figure 7.30.
4 Guidance will be required from an experienced person.
5 The question suggests a lot of bleeding and the necessary action to stop it is by applying direct pressure to the wound. See chapter notes on first-aid treatment.
6 A diagram of this transformer is shown in Figure 7.31. Apart from both windings being completely isolated from each other, the core is earthed independently.
7 Firstly, they are both methods of extra-low voltage. SELV is completely isolated from earth (i.e. an earth-free supply) and a person touching live conductors is safe from electric shock since there is no return path for any current. FELV differs from SELV by being earthed at one point. It is permissible to use a SELV circuit in a bathroom, and a bell can be supplied from a FELV circuit.
8 Two methods are shown in Figure 7.32.

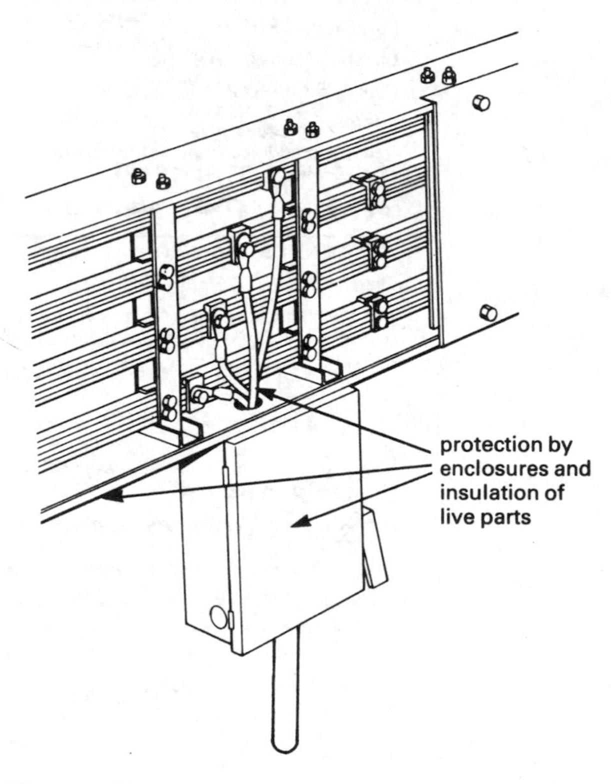

Figure 7.32

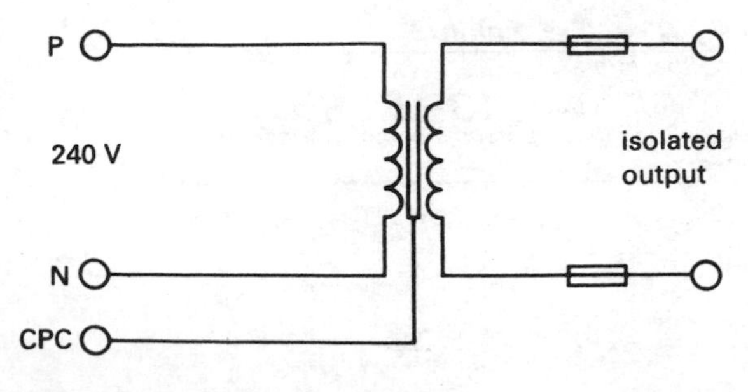

Figure 7.31 *Typical small factory wiring distribution*

9 Automatic disconnection of circuit protective devices and/or earth leakage trips can only occur if the means of earthing is properly carried out. The maximum impedance values for protective devices are given in the Tables of the IEE Wiring Regulations.
10 Fixed circuit equipment is less vulnerable to misuse than equipment connected to socket outlets via flexible leads and risk of electric shock less likely.

Exercise 3

(1) (a) The question does not provide the ceiling to switch heights but if this was to be assumed as at least 1 m then approximately 180 m of conduit is required (60 lengths), allowing for wastage.

(b) There will be two circuit protective conductors in the power conduits and this implies that six cables will be required in most places, i.e. assuming the socket outlets are wired as a ring. *Note:* The question is from a CGLI 1986 Assignment Paper (Rewiring of a flat).

2 The solution to this question is shown in Figure 7.33.

3 See Figures 3.2 and 3.3.

4 Information on tower scaffolds can be obtained from the HSE, Guidance Notes GS42—this can be found in your college library.

5 (a) This method of time-keeping lends itself favourably to employers who have a large workforce, where accurate time-keeping is essential to encourage punctuality. The system often preferred is the punch card method which takes a relatively short time to carry out and provides the employer with immediate information of employees' time-keeping to work out wages.

In service industries like the electrical contracting industry, the punch card system might be difficult to operate, mainly because of contracts worked away from the office and overtime working. In such circumstances, a good charge-hand or foreman is all that is required to organise a small workforce.

(b) These are taken at set times laid down by the employer. Meal breaks, including washing time, are generally of one hour duration and are unpaid, whereas tea-breaks, which are usually ten/fifteen minute periods in the morning and afternoon, are paid by the employer.

It is in the interest of the employees not to exceed the times granted by their employers as this will only result in lost production time and lost profit for the employer and eventual unemployment for the employees.

(c) These facilities will be found on the premises of the work place. They may be provided by the employer in a site hut or cabin or provided by the main builder, contractor or client as a shared facility. Such facilities will enable the employees to maintain a reasonable standard of hygiene at the place of work

(d) These are usually arranged by the main builder of a large contract to discuss its progress/delay/alterations, etc. The meetings bring together interested parties such as electricians, plumbers and carpenters, etc. Generally it is the charge-hands and foremen of the respective trades who become involved and who have the opportunity to discuss their problems. Such meetings should aim to improve the relations between the different contractors and allow a free flow of work for ensuing weeks until the next meeting.

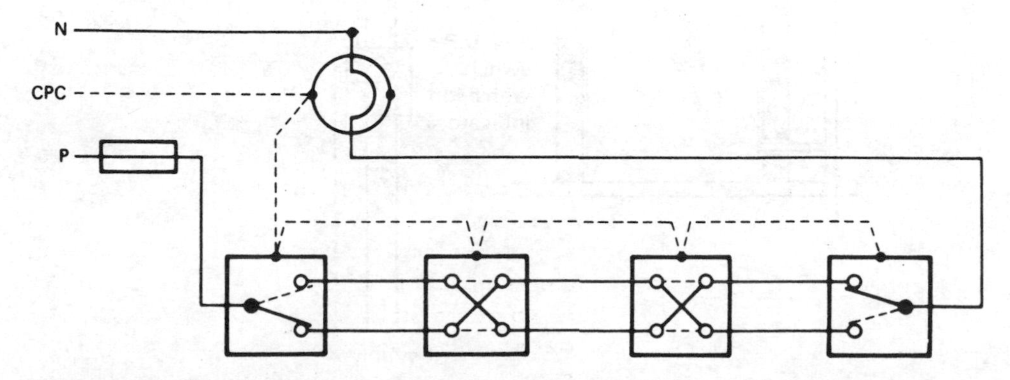

Figure 7.33

It is good practice to keep a record of the minutes of each meeting and for them to be agreed and signed before each new meeting commences.

6 (a) Check the delivery note with a copy of the actual order; check for any damage or missing items. Do not sign for any missing items.
(b) Keep delicate items (glass fittings and surface fittings) completely separate from other site items that could possibly cause damage. Use additional hut or room.
(c) This will contain telephone numbers, dates of meetings and visits, etc.

7 A typical circuit diagram is shown in Figure 4.4.

8 (a) (i) The conduit factor is 213 and since 1.5 mm^2 has a factor of 22, then 231/22 = 10.5 (approx. 11).
(ii) The conduit factor is 204 and since 2.5 mm^2 has a factor of 30, then 204/30 = 6.8 (approx. 7).
(b) The overall cable factor is 372. Only 32 mm conduit can be used, having a factor of 600.

9 Some of these points have been given in the text, but see HSE Guidance Notes GS33.

10 Consult manufacturers' catalogues. For example,

(a) Crabtree 2214 QG.
(b) MK 3591 ALM.
(c) Crabree 708/2
(d) Two 50/05; one 50/15; one 50/20; two 50/30 and two 743 blank plates.
(e) Thorn PPQ 675 plus PPC6.

Exercise 4

1 Since three phases are available, then two cooking appliances will be connected to each phase (with one phase lightly loaded). The largest appliance takes a current of
$I = P/V = 15\,000/240 = 62.5$ A and the second-largest takes a current of
$I = P/V = 10\,000/240 = 41.67$ A. Thus each phase requires a current of
$62.5 + (0.8 \times 41.67) = 95.84$ A.

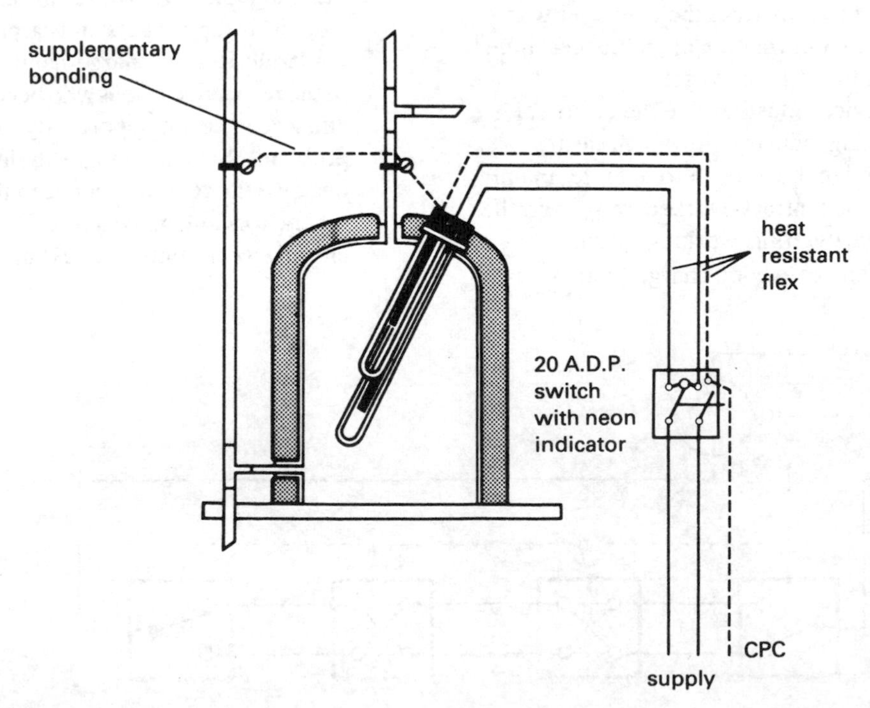

Figure 7.34 *Immersion heater final circuit showing supplementary bonding*

2 See page 124. The term sensitivity refers to the tripping current of an r.c.d. If it is low (say 20 mA) it will have a high sensitivity; if it is high (say 300 mA) it has low sensitivity.
3 See Fig. 7.34.
4 (a) See page 62 of Chapter 4.
(b) $I_f = U_o/Z_s = 240/1.14 = 210.5$ A. Since $t = 0.4$ s (for socket outlets) and $k = 115$ (for copper), then $S = 1.15$ mm^2.
5 (i) Expressed as a percentage, it takes into account some equipment that will never be fully utilised while other current-using equipment is being used. It will enable the designer of an installation to economically assess the sizes of main cables and main switchgear for the installation.
(ii) This is a method of estimating the loading of current-using equipment; for example a GLS lighting point will have a minimum connected load of 100 W despite several situations where a lower wattage lamp may be used. Similarly, a fluorescent luminaire or other discharge luminaire will have its operating power multiplied by a factor of 1.8 to compensate for the power factor and harmonic currents set up in the ballast. A 2 A socket outlet has an assumed current demand of 0.5 A. For a cooking appliance, see the assessment made on page 60.

6 5 A is 0.015 s; 15 A is 0.9 s; 20 A is 50 s and the 30 A is 1100 s. Apart from a quicker response time at the lower end, they indicate proper discrimination with smaller size fuses rupturing first.

7

Protective device	*Advantages*	*Disadvantages*
BS 3036 fuse	Cheap initial cost No mechanical movement	Lack of discrimination Incorrect fuse element
BS 88 fuse	Reliable operation Good discrimination	Expensive Slow to restore supply
BS 3871 T1 m.c.b.	Circuit is easily seen to be off Tamperproof	Has moving parts Need for regular tests

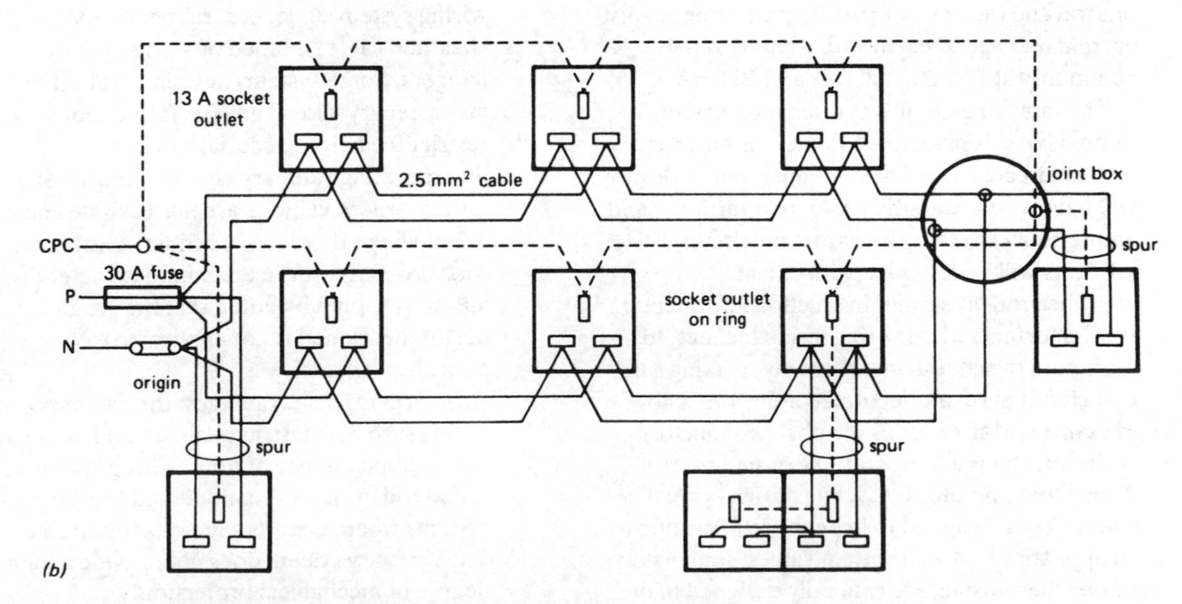

Figure 7.35 *Ring final circuit feeding 13 A socket outlets*

8 See page 67.
9 See pages 67–70.
10 $I_b = 25$ A; I_n is 30 A; $I_z = 42.66$ A; 10 mm^2; $I_F = 245.6$ A and $S = 1.17$ mm^2. The installed protective conductor is larger than the calculated value and therefore satisfies the thermal constraints.

Exercise 5

1 See Figure 7.35.
2 A metal clad consumer unit is a factory-built assembly (see 'Definitions', IEE Wiring Regulations, page 9) and for indoor use, designed to the specification of BS 5486, Part 13 (1979). The two main features of the consumer unit will be miniature circuit breakers and main switch which often takes the form of a double-pole earth leakage circuit breaker of the residual current type (i.e. r.c.d.), rated between 30 A and 100 A.
The miniature circuit breakers are to BS 3871 specification and come in current ratings of 5 A, 10 A, 15 A, 20 A, 30 A, 40 A and 50 A. These devices are arranged in line with each other with a common busbar linking the lower m.c.b. terminals and the outgoing phase connection of the r.c.d. The r.c.d. also has an outgoing multi-terminal neutral and the device provides protection against current leakage to earth with tripping values commonly at 30 mA, 100 mA and 300 mA.
The interior assembly is often mounted on a removable sub-plate to allow the consumer unit to be fixed in position. The consumer unit enclosure will have a conveniently positioned earth bar and ample knockout entry holes for final circuit wiring. All units will be provided with circuit identification labels and instruction information. It is important, when wiring inside the unit, to terminate the conductors correctly, making sure that circuit earth and neutral conductors follow the same wiring order as the phase conductors.
3 A thermostat is a temperature-sensing control device that operates by cycling during normal use and which is designed to keep the temperature of an appliance between certain values. Immersion heater thermostats are generally designed to the specifications of BS 3955 and the tests prescribed therein to see if they have insufficient self-heat to cycle when their temperature-sensing elements are maintained at any suitable constant temperature.
The type often used for controlling temperature of domestic water heaters is made of brass and operates a single-pole micro-gap switch. Figure 7.36 illustrates this type of thermostat, where it will be seen that the brass tube and invar steel rod (non-expanding rod) are joined at one end. This is done so that when the temperature rises in the heated water, the expansion of the brass tube reduces the pressure on the pressure block and the contacts separate. The small magnet allows the mechanism to have a snap action effect to avoid unnecessary sparking and radio interference. The temperature scale and adjusting knob are fitted in the head of a moulded plastic cover. It should be pointed out that in hard water areas, the thermostat setting should not be higher than 60–65°C.
4 (a) This system is often used as an alternative to burying a wiring system in the ground, which could prove costly. It is ideal for temporary supplies and supplies needed for outbuildings as found in agricultural and horticultural installations. The wire used is galvanized steel wire which is strained tight. From this the wiring system cable, e.g. rubber or PVC sheathed cable, is taped or suspended by hangers. Some systems have integral cable and catenary wire. The IEE Regulations restrict the height of aerial cables incorporating catenary wire to 3.5 m above ground where vehicles are inaccessible and 5.2 m where they are accessible, apart from road crossings where the minimum height is 5.8 m. It should be noted that the 3.5 m height mentioned is not applicable to agricultural premises.
In terms of disadvantages, the catenary wiring system is restricted in use and has not the flexibility of use of other wiring systems found today, such as conduit and trunking systems where circuits can be easily altered. The catenary system does not provide a high degree of mechanical protection.
(b) In this system, the immediate advantage is

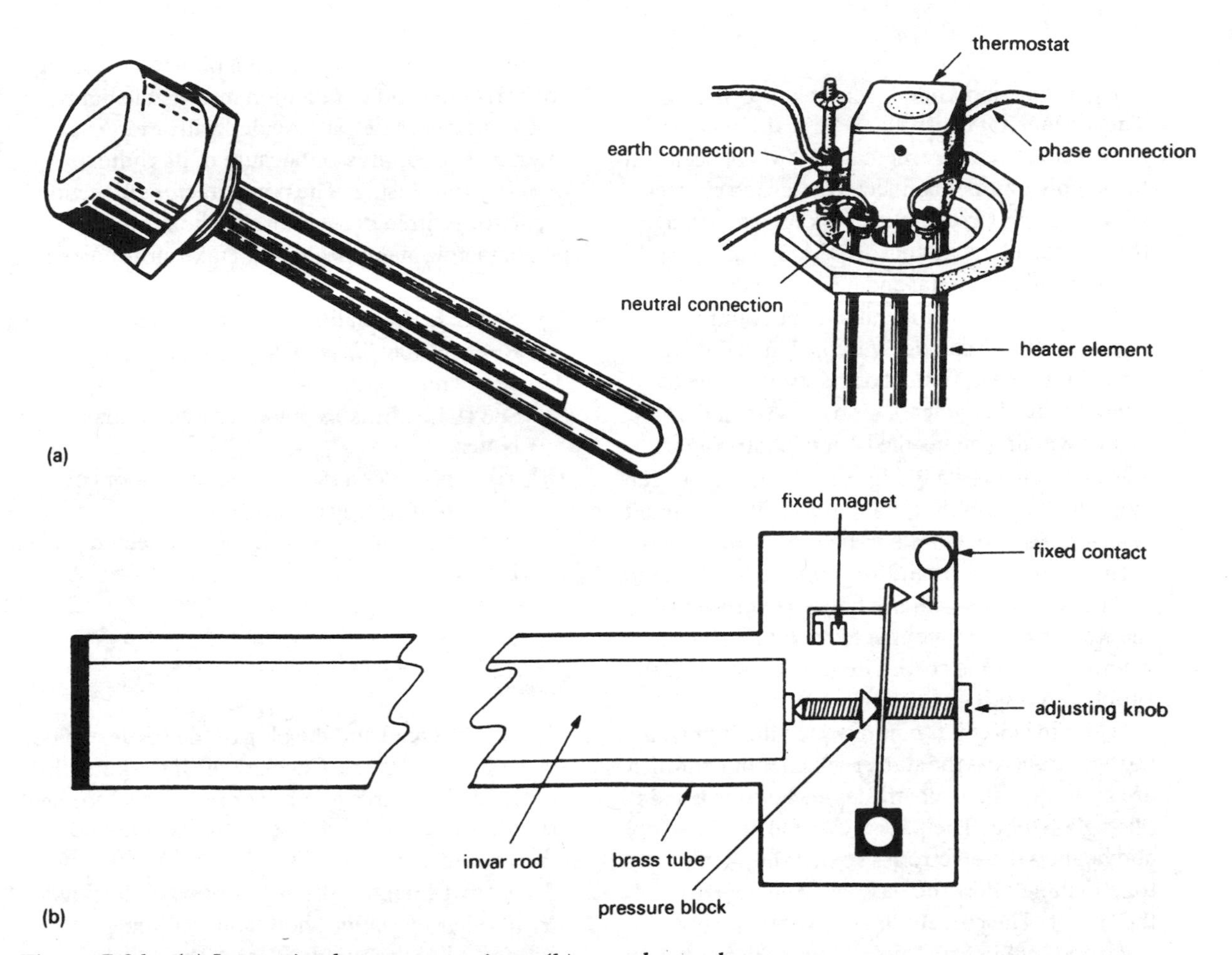

Figure 7.36 *(a) Immersion heater connections; (b) water heater thermostat*

found by the sheath being used as a return conductor. Basically, the wiring system used is m.i.m.s. cable since its outer sheath is copper which provides an ideal PEN conductor. Sealing pots which contain an earthing tail are used for this purpose. While the system has the advantage of not requiring any neutral conductors contained within the cable itself, its use as a wiring system is somewhat restricted to installations not connected directly to the public supply. Furthermore, where m.i.m.s. is used as the wiring system, the sheath must not have a c.s.a. of less than 4 mm^2 and its resistance must never be more than any of its internal conductors. Other conditions for its use are given in the IEE Wiring Regulations, Regs. 546–1 to 546–8.

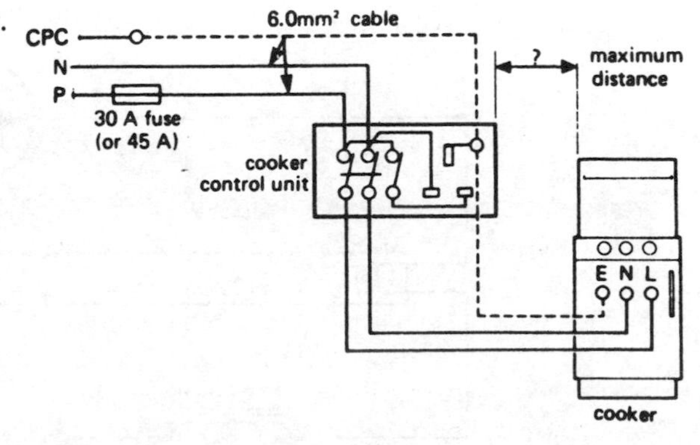

Figure 7.37

5 (a) See Figure 7.37
(b) 2 m, recommended (see Reg. 476–03–04).

6 The simmerstat controller is basically a variable heat switch or energy regulator. It will be seen that the supply phase is connected to a fixed contact while a moving contact conveys current through the heating element and also through a compensating bimetal and heater winding. A control cam adjusts the initial temperature required and its contact with the bimetallic strip only allows the lower leaf of the strip to move away from the fixed contact. This occurs when the heater winding bends the bimetallic strip apart.

7 The circuit is shown in Figure 7.38. It will be seen from the diagram that the fluorescent tube circuit consists of a choke or lamp ballast, power factor correction capacitor and starter switch. The lamp ballast is in series with the fluorescent tube while the starter switch is connected across the tube electrodes and the correction capacitor across the supply terminals of the lamp circuit.

When the circuit is connected to the supply, a p.d. occurs across the starter switch contacts which are bimetallic strip electrodes enclosed in a neon filled glass tube. The p.d. causes the neon to glow and heat up the electrodes so that they come together and allow the passage of current around the circuit. The current heats up both lamp electrodes but because no p.d. exists in the starter switch at this moment, its bimetallic contacts cool down and spring apart. The breaking of these contacts causes the choke to produce a momentary high voltage that is sufficient to strike an arc in the fluorescent tube and its internal phosphor coating converts ultra-violet radiation into visible light.

The starter switch is provided with a radio interference suppressor because of its contacts opening and closing. The power factor correction capacitor is fitted because of the choke's poor power factor, making the circuit take more current than necessary.

8 (a) See IEE Regulations' Index, particularly Reg. 130–06–02 and Chapter 55.
(b) See Figure 7.39

9 (a) See IEE Wiring Regulations and Guidance Notes.
(b) (i) provides a supply to several floors of multistorey buildings
(ii) floor supplies are easily connected.

10 See Figure 7.40.

Exercise 6

1 Briefly, the lead and the plug of the kettle can be checked for damage. Check to see if the kettle has a British Standards number or recognised foreign standard. Make the necessary tests explained in this chapter.

2 See end of Chapter 6 on inspection and testing.

3 Your diagram should show a buzzer being connected to the lampholder via the switch controlling it, and connection should be on the centre contact when the switch is closed.

4 The ohmmeter can be used for these three operations, testing each core to the sheath and

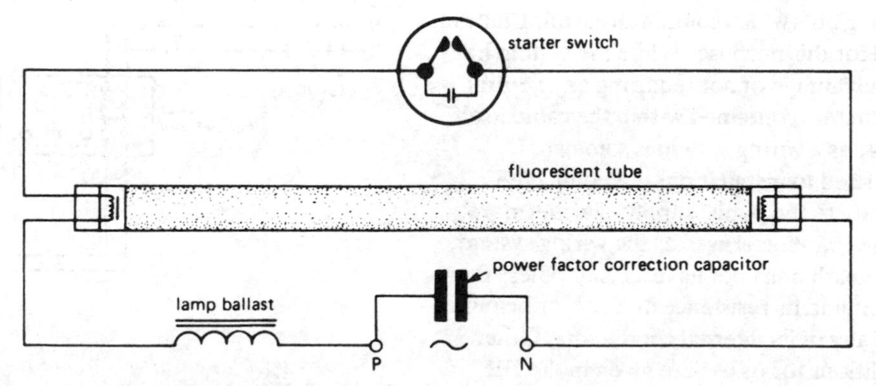

Figure 7.38 *Fluorescent lamp circuit*

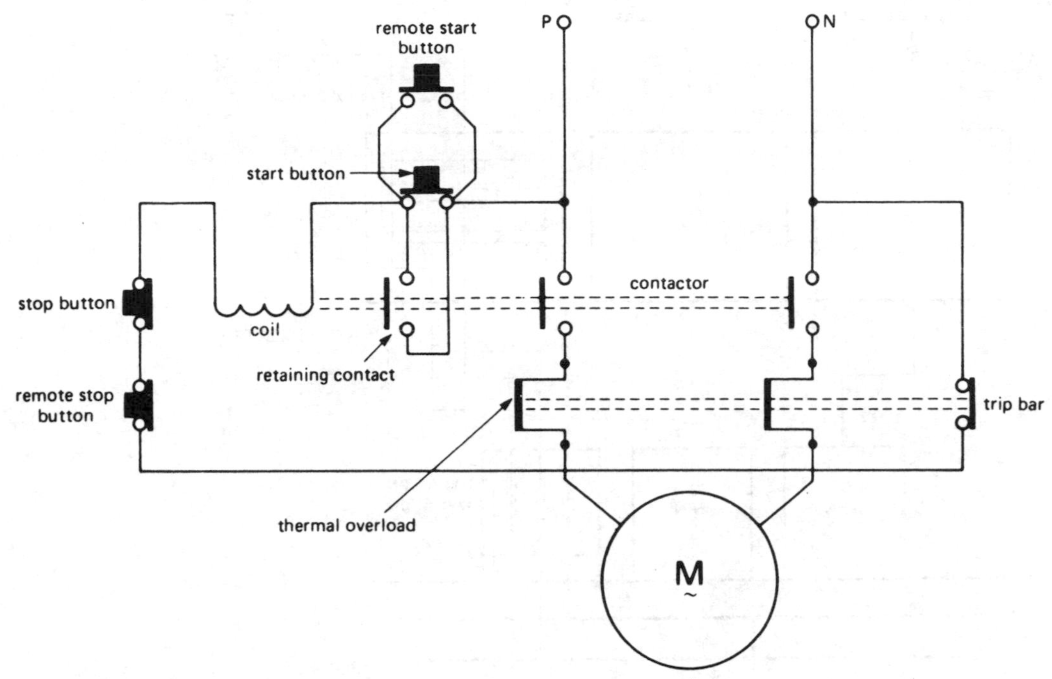

Figure 7.39 *Direct-on-line contact starter for single- phase motor*

then testing for insulation resistance between them and also to earth.

5 The motor may or may not have its three stator windings on view. If it does, then they need to be separated and tested individually for continuity. All three readings should be the same. For insulation resistance, the ends can be joined together and tested to earth, as well as making a test between each other. A reading above 0.5Ω is regarded as being satisfactory for insulation resistance.

6 Reference should be made to the tables of the IEE Regulations.
   (a) 5.58Ω for a 15 A BS 3036 fuse—disconnection is 5 s
   (b) 1.92Ω for a 32 A BS 88 fuse—disconnection is 5 s
   (c) 2.0Ω for Type 1; 1.14Ω for Type 2; and 0.8Ω for Type 3 m.c.b.s. to BS 3871—disconnection is 0.4 s.

7 If the test was done at the main point, all the section results would be an equivalent parallel value, i.e. 0.78 MΩ.

8 See Chapter 6 on inspection and testing, as well as Part 7, IEE Regulations, Appendix 6.

9 Reference needs to be made to Reg. 413–02–06, i.e. the product of the r.c.d. tripping current and earth fault loop impedance must not exceed 50. Since the device trips at 30 mA, then the maximum impedance is:
$Z(\text{max.}) = 50/0.03 = 1667\Omega$.

Note: Any high value of earth electrode resistance will not allow circuit protective devices to operate under earth fault conditions.

10 Reference should be made to HSE Guidance Notes GS 38. It is important that test lamps are regularly proved before and after use, that the lamp has its correct leads, probes and connectors, and that it is suitable for the working voltage. The lamp would be dim on 240 V and bright on 415 V.

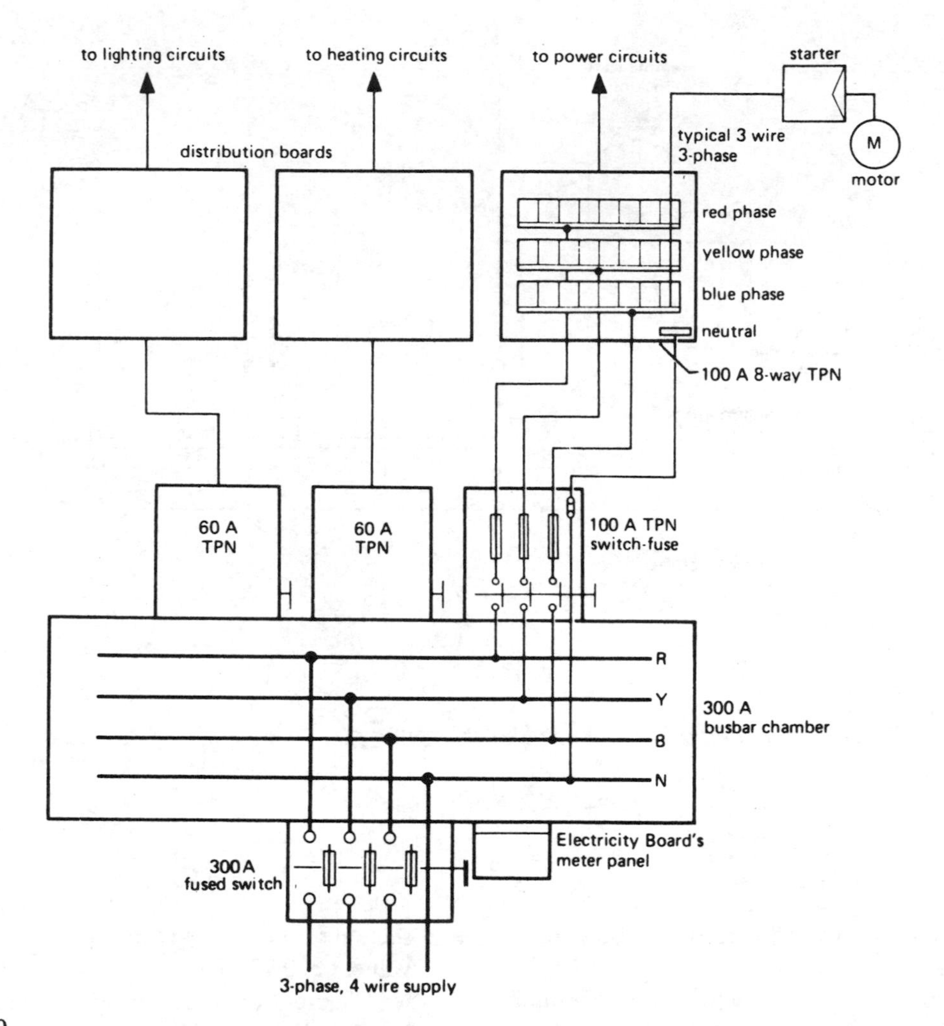

Figure 7.40

Answers to multiple-choice question paper

Part I Certificate

1 (c)	11 (c)	21 (c)	31 (c)	41 (a)
2 (b)	12 (b)	22 (c)	32 (c)	42 (d)
3 (a)	13 (a)	23 (b)	33 (b)	43 (a)
4 (d)	14 (b)	24 (b)	34 (c)	44 (a)
5 (d)	15 (b)	25 (c)	35 (d)	45 (b)
6 (b)	16 (c)	26 (d)	36 (b)	46 (a)
7 (c)	17 (a)	27 (d)	37 (b)	47 (c)
8 (b)	18 (b)	28 (d)	38 (a)	48 (b)
9 (c)	19 (d)	29 (a)	39 (d)	49 (c)
10 (d)	20 (a)	30 (b)	40 (c)	50 (b)

Part II Certificate

1 (c)	11 (b)	21 (d)	31 (b)	41 (c)
2 (b)	12 (a)	22 (a)	32 (c)	42 (b)
3 (a)	13 (d)	23 (d)	33 (c)	43 (d)
4 (c)	14 (c)	24 (c)	34 (a)	44 (d)
5 (d)	15 (b)	25 (b)	35 (d)	45 (a)
6 (c)	16 (b)	26 (a)	36 (b)	46 (a)
7 (b)	17 (d)	27 (a)	37 (d)	47 (c)
8 (b)	18 (b)	28 (a)	38 (a)	48 (b)
9 (c)	19 (b)	29 (a)	39 (a)	49 (c)
10 (d)	20 (b)	30 (d)	40 (c)	50 (a)

Index